SUPERBUGS AND SUPERDRUGS:
A HISTORY OF MRSA

The transcript of a Witness Seminar held by the Wellcome Trust Centre for the History of Medicine at UCL, London, on 11 July 2006

Edited by L A Reynolds and E M Tansey

Volume 32   2008

First published by the Wellcome Trust Centre
for the History of Medicine at UCL, 2008

The Wellcome Trust Centre for the History of Medicine
at UCL is funded by the Wellcome Trust, which is
a registered charity, no. 210183.

ISBN 978 085484 114 1

All volumes are freely available online following the links to Publications/Wellcome Witnesses at
www.ucl.ac.uk/histmed

Technology Transfer in Britain: The case of monoclonal antibodies; Self and Non-Self: A history of autoimmunity; Endogenous Opiates; The Committee on Safety of Drugs • Making the Human Body Transparent: The impact of NMR and MRI; Research in General Practice; Drugs in Psychiatric Practice; The MRC Common Cold Unit • Early Heart Transplant Surgery in the UK • Haemophilia: Recent history of clinical management • Looking at the Unborn: Historical aspects of obstetric ultrasound • Post Penicillin Antibiotics: From acceptance to resistance? • Clinical Research in Britain, 1950–1980 • Intestinal Absorption • Origins of Neonatal Intensive Care in the UK • British Contributions to Medical Research and Education in Africa after the Second World War • Childhood Asthma and Beyond • Maternal Care • Population-based Research in South Wales: The MRC Pneumoconiosis Research Unit and the MRC Epidemiology Unit • Peptic Ulcer: Rise and fall • Leukaemia • The MRC Applied Psychology Unit • Genetic Testing • Foot and Mouth Disease: The 1967 outbreak and its aftermath • Environmental Toxicology: The legacy of Silent Spring • Cystic Fibrosis • Innovation in Pain Management • The Rhesus Factor and Disease Prevention • The Recent History of Platelets in Thrombosis and Other Disorders • Short-course Chemotherapy for Tuberculosis • Prenatal Corticosteroids for Reducing Morbidity and Mortality after Preterm Birth • Public Health in the 1980s and 1990s: Decline and rise? • Cholesterol, Atherosclerosis and Coronary Disease in the UK, 1950–2000 • Development of Physics Applied to Medicine in the UK, 1945–90 • The Early Development of Total Hip Replacement • The Discovery, Use and Impact of Platinum Salts as Chemotherapy Agents for Cancer • Medical Ethics Education in Britain, 1963–1993 • Superbugs and Superdrugs: A history of MRSA

CONTENTS

ILLUSTRATIONS AND CREDITS

WITNESS SEMINARS:
MEETINGS AND PUBLICATIONS[1]

In 1990 the Wellcome Trust created a History of Twentieth Century Medicine Group, associated with the Academic Unit of the Wellcome Institute for the History of Medicine, to bring together clinicians, scientists, historians and others interested in contemporary medical history. Among a number of other initiatives the format of Witness Seminars, used by the Institute of Contemporary British History to address issues of recent political history, was adopted, to promote interaction between these different groups, to emphasize the potential benefits of working jointly, and to encourage the creation and deposit of archival sources for present and future use. In June 1999 the Governors of the Wellcome Trust decided that it would be appropriate for the Academic Unit to enjoy a more formal academic affiliation and turned the Unit into the Wellcome Trust Centre for the History of Medicine at UCL from 1 October 2000. The Wellcome Trust continues to fund the Witness Seminar programme via its support for the Centre.

The Witness Seminar is a particularly specialized form of oral history, where several people associated with a particular set of circumstances or events are invited to come together to discuss, debate, and agree or disagree about their memories. To date, the History of Twentieth Century Medicine Group has held more than 50 such meetings, most of which have been published, as listed on pages xi–xix.

Subjects are usually proposed by, or through, members of the Programme Committee of the Group, which includes professional historians of medicine, practising scientists and clinicians, and once an appropriate topic has been agreed, suitable participants are identified and invited. This inevitably leads to further contacts, and more suggestions of people to invite. As the organization of the meeting progresses, a flexible outline plan for the meeting is devised, usually with assistance from the meeting's chairman, and some participants are invited to 'set the ball rolling' on particular themes, by speaking for a short period to initiate and stimulate further discussion.

---

[1] The following text also appears in the 'Introduction' to recent volumes of *Wellcome Witnesses to Twentieth Century Medicine* published by the Wellcome Trust and the Wellcome Trust Centre for the History of Medicine at UCL.

Each meeting is fully recorded, the tapes are transcribed and the unedited transcript is immediately sent to every participant. Each is asked to check his or her own contributions and to provide brief biographical details. The editors turn the transcript into readable text, and participants' minor corrections and comments are incorporated into that text, while biographical and bibliographical details are added as footnotes, as are more substantial comments and additional material provided by participants. The final scripts are then sent to every contributor, accompanied by forms assigning copyright to the Wellcome Trust. Copies of all additional correspondence received during the editorial process are deposited with the records of each meeting in Archives and Manuscripts, Wellcome Library, London.

As with all our meetings, we hope that even if the precise details of some of the technical sections are not clear to the non-specialist, the sense and significance of the events will be understandable. Our aim is for the volumes that emerge from these meetings to inform those with a general interest in the history of modern medicine and medical science; to provide historians with new insights, fresh material for study, and further themes for research; and to emphasize to the participants that events of the recent past, of their own working lives, are of proper and necessary concern to historians.

ACKNOWLEDGEMENTS

'Superbugs and superdrugs: the history of MRSA' was suggested as a suitable topic for a Witness Seminar by Professor Gordon Stewart, who assisted us in planning the meeting. We are very grateful to him for his input and to Dr Robert Bud for his excellent chairing of the occasion. We are particularly grateful to Professor David Greenwood for writing such a useful Introduction to these published proceedings. Our additional thanks go to Professor Gordon Stewart and Dr Angela Kearns for their help preparing the appendices for publication; Professor David Greenwood who read through earlier drafts of the transcript, and offered helpful comments and advice. We thank the contributors for their help with the Glossary, in particular, Professors Mark Casewell, David Greenwood, Jeremy Hamilton-Miller, Ian Phillips, Brian Spratt and Gordon Stewart; and for help with photographs we thank Professors Graham Ayliffe and David Greenwood, Dr Georgia Duckworth, Dr Angela Kearns and Dr Bill Newsom. For permission to reproduce the images included here, we thank the *Chemist and Druggist, Clinical Infectious Diseases,* the Health Protection Agency and the Wellcome Trust; and Ron Alfa for editorial assistance.

As with all our meetings, we depend a great deal on our colleagues at the Wellcome Trust to ensure their smooth running: the Audiovisual Department, and the Medical Photographic Library; Mr Akio Morishima, who has supervised the design and production of this volume; our indexer, Ms Liza Furnival; and our readers, Ms Fiona Plowman, Mrs Sarah Beanland and Mr Simon Reynolds. Mrs Jaqui Carter is our transcriber, and Mrs Wendy Kutner and Dr Daphne Christie assist us in running the meetings. Finally we thank the Wellcome Trust for supporting this programme.

Tilli Tansey

Lois Reynolds

Wellcome Trust Centre for the History of Medicine at UCL

HISTORY OF TWENTIETH CENTURY MEDICINE WITNESS SEMINARS, 1993–2008

PUBLISHED MEETINGS

'…Few books are so intellectually stimulating or uplifting'.
Journal of the Royal Society of Medicine (1999) **92:** 206–8,
review of vols 1 and 2

*'…This is oral history at its best…all the volumes make compulsive reading…they
are, primarily, important historical records'.*
British Medical Journal (2002) **325:** 1119, review of the series

Technology transfer in Britain: The case of monoclonal antibodies
Self and non-self: A history of autoimmunity
Endogenous opiates
The Committee on Safety of Drugs
Tansey E M, Catterall P P, Christie D A, Willhoft S V, Reynolds L A. (eds)
(1997) *Wellcome Witnesses to Twentieth Century Medicine.* Volume 1. London:
The Wellcome Trust, 135pp. ISBN 1 869835 79 4

Making the human body transparent: The impact of NMR and MRI
Research in general practice
Drugs in psychiatric practice
The MRC Common Cold Unit
Tansey E M, Christie D A, Reynolds L A. (eds) (1998) *Wellcome
Witnesses to Twentieth Century Medicine.* Volume 2. London: The Wellcome
Trust, 282pp. ISBN 1 869835 39 5

Early heart transplant surgery in the UK
Tansey E M, Reynolds L A. (eds) (1999) *Wellcome Witnesses to
Twentieth Century Medicine.* Volume 3. London: The Wellcome Trust, 72pp.
ISBN 1 841290 07 6

Haemophilia: Recent history of clinical management
Tansey E M, Christie D A. (eds) (1999) *Wellcome Witnesses to
Twentieth Century Medicine.* Volume 4. London: The Wellcome Trust, 90pp.
ISBN 1 841290 08 4

Looking at the unborn: Historical aspects of obstetric ultrasound
Tansey E M, Christie D A. (eds) (2000) *Wellcome Witnesses to
Twentieth Century Medicine.* Volume 5. London: The Wellcome Trust, 80pp.
ISBN 1 841290 11 4

Post penicillin antibiotics: From acceptance to resistance?
Tansey E M, Reynolds L A. (eds) (2000) *Wellcome Witnesses to Twentieth Century Medicine.* Volume 6. London: The Wellcome Trust, 71pp. ISBN 1 841290 12 2

Clinical research in Britain, 1950–1980
Reynolds L A, Tansey E M. (eds) (2000) *Wellcome Witnesses to Twentieth Century Medicine.* Volume 7. London: The Wellcome Trust, 74pp. ISBN 1 841290 16 5

Intestinal absorption
Christie D A, Tansey E M. (eds) (2000) *Wellcome Witnesses to Twentieth Century Medicine.* Volume 8. London: The Wellcome Trust, 81pp. ISBN 1 841290 17 3

Neonatal intensive care
Christie D A, Tansey E M. (eds) (2001) *Wellcome Witnesses to Twentieth Century Medicine.* Volume 9. London: The Wellcome Trust Centre for the History of Medicine at UCL, 84pp. ISBN 0 854840 76 1

British contributions to medical research and education in Africa after the Second World War
Reynolds L A, Tansey E M. (eds) (2001) *Wellcome Witnesses to Twentieth Century Medicine.* Volume 10. London: The Wellcome Trust Centre for the History of Medicine at UCL, 93pp. ISBN 0 854840 77 X

Childhood asthma and beyond
Reynolds L A, Tansey E M. (eds) (2001) *Wellcome Witnesses to Twentieth Century Medicine.* Volume 11. London: The Wellcome Trust Centre for the History of Medicine at UCL, 74pp. ISBN 0 854840 78 8

Maternal care
Christie D A, Tansey E M. (eds) (2001) *Wellcome Witnesses to Twentieth Century Medicine.* Volume 12. London: The Wellcome Trust Centre for the History of Medicine at UCL, 88pp. ISBN 0 854840 79 6

Population-based research in south Wales: The MRC Pneumoconiosis Research Unit and the MRC Epidemiology Unit
Ness A R, Reynolds L A, Tansey E M. (eds) (2002) *Wellcome Witnesses to Twentieth Century Medicine.* Volume 13. London: The Wellcome Trust Centre for the History of Medicine at UCL, 74pp. ISBN 0 854840 81 8

Peptic ulcer: Rise and fall

Christie D A, Tansey E M. (eds) (2002) *Wellcome Witnesses to Twentieth Century Medicine.* Volume 14. London: The Wellcome Trust Centre for the History of Medicine at UCL, 143pp. ISBN 0 854840 84 2

Leukaemia

Christie D A, Tansey E M. (eds) (2003) *Wellcome Witnesses to Twentieth Century Medicine.* Volume 15. London: The Wellcome Trust Centre for the History of Medicine at UCL, 86pp. ISBN 0 85484 087 7

The MRC Applied Psychology Unit

Reynolds L A, Tansey E M. (eds) (2003) *Wellcome Witnesses to Twentieth Century Medicine.* Volume 16. London: The Wellcome Trust Centre for the History of Medicine at UCL, 94pp. ISBN 0 85484 088 5

Genetic testing

Christie D A, Tansey E M. (eds) (2003) *Wellcome Witnesses to Twentieth Century Medicine.* Volume 17. London: The Wellcome Trust Centre for the History of Medicine at UCL, 130pp. ISBN 0 85484 094 X

Foot and mouth disease: The 1967 outbreak and its aftermath

Reynolds L A, Tansey E M. (eds) (2003) *Wellcome Witnesses to Twentieth Century Medicine.* Volume 18. London: The Wellcome Trust Centre for the History of Medicine at UCL, 114pp. ISBN 0 85484 096 6

Environmental toxicology: The legacy of *Silent Spring*

Christie D A, Tansey E M. (eds) (2004) *Wellcome Witnesses to Twentieth Century Medicine.* Volume 19. London: The Wellcome Trust Centre for the History of Medicine at UCL, 132pp. ISBN 0 85484 091 5

Cystic fibrosis

Christie D A, Tansey E M. (eds) (2004) *Wellcome Witnesses to Twentieth Century Medicine.* Volume 20. London: The Wellcome Trust Centre for the History of Medicine at UCL, 120pp. ISBN 0 85484 086 9

Innovation in pain management

Reynolds L A, Tansey E M. (eds) (2004) *Wellcome Witnesses to Twentieth Century Medicine.* Volume 21. London: The Wellcome Trust Centre for the History of Medicine at UCL, 125pp. ISBN 0 85484 097 4

The Rhesus factor and disease prevention
Zallen D T, Christie D A, Tansey E M. (eds) (2004) *Wellcome Witnesses to Twentieth Century Medicine.* Volume 22. London: The Wellcome Trust Centre for the History of Medicine at UCL, 98pp. ISBN 0 85484 099 0

The recent history of platelets in thrombosis and other disorders
Reynolds L A, Tansey E M. (eds) (2005) *Wellcome Witnesses to Twentieth Century Medicine.* Volume 23. London: The Wellcome Trust Centre for the History of Medicine at UCL, 186pp. ISBN 0 85484 103 2

Short-course chemotherapy for tuberculosis
Christie D A, Tansey E M. (eds) (2005) *Wellcome Witnesses to Twentieth Century Medicine.* Volume 24. London: The Wellcome Trust Centre for the History of Medicine at UCL, 120pp. ISBN 0 85484 104 0

Prenatal corticosteroids for reducing morbidity and mortality after preterm birth
Reynolds L A, Tansey E M. (eds) (2005) *Wellcome Witnesses to Twentieth Century Medicine.* Volume 25. London: The Wellcome Trust Centre for the History of Medicine at UCL, 154pp. ISBN 0 85484 102 4

Public health in the 1980s and 1990s: Decline and rise?
Berridge V, Christie D A, Tansey E M. (eds) (2006) *Wellcome Witnesses to Twentieth Century Medicine.* Volume 26. London: The Wellcome Trust Centre for the History of Medicine at UCL, 101pp. ISBN 0 85484 106 7

Cholesterol, atherosclerosis and coronary disease in the UK, 1950–2000
Reynolds L A, Tansey E M. (eds) (2006) *Wellcome Witnesses to Twentieth Century Medicine.* Volume 27. London: The Wellcome Trust Centre for the History of Medicine at UCL, 164pp. ISBN 0 85484 107 5

Development of physics applied to medicine in the UK, 1945–90
Christie D A, Tansey E M. (eds) (2006) *Wellcome Witnesses to Twentieth Century Medicine.* Volume 28. The Wellcome Trust Centre for the History of Medicine at UCL, 141pp. ISBN 0 85484 108 3

Early development of total hip replacement
Reynolds L A, Tansey E M. (eds) (2007) *Wellcome Witnesses to Twentieth Century Medicine.* Volume 29. London: The Wellcome Trust Centre for the History of Medicine at UCL, 198pp. ISBN 978 085484 111 0

The discovery, use and impact of platinum salts as chemotherapy agents for cancer
Christie D A, Tansey E M. (eds) (2007) *Wellcome Witnesses to Twentieth Century Medicine.* Volume 30. London: The Wellcome Trust Centre for the History of Medicine at UCL, 142pp. ISBN 978 085484 112 7

Medical Ethics Education in Britain, 1963–93
Reynolds L A, Tansey E M. (eds) (2007) *Wellcome Witnesses to Twentieth Century Medicine.* Volume 31. London: The Wellcome Trust Centre for the History of Medicine at UCL, 241pp. ISBN 978 085484 113 4

Superbugs and superdrugs: A history of MRSA
Reynolds L A, Tansey E M. (eds) (2008) *Wellcome Witnesses to Twentieth Century Medicine.* Volume 32. London: The Wellcome Trust Centre for the History of Medicine at UCL, (this volume). ISBN 978 085484 114 1

Clinical pharmacology in the UK, *c.* 1950–2000
Christie D A, Tansey E M. (eds) (2008) *Wellcome Witnesses to Twentieth Century Medicine.* Volume 33. London: The Wellcome Trust Centre for the History of Medicine at UCL, (in press). ISBN 978 085484117 2

The resurgence of breast-feeding, 1975–2000
Christie D A, Tansey E M. (eds) (2008) *Wellcome Witnesses to Twentieth Century Medicine.* Volume 34. London: The Wellcome Trust Centre for the History of Medicine at UCL, (in press). ISBN 978 085484118 9

Hard copies of volumes 1–20 are now available for free, while stocks last. We would be happy to send complete sets to libraries in developing or restructuring countries. Available from Dr Carole Reeves at: *c.reeves@ucl.ac.uk*

All volumes are freely available online at www.ucl.ac.uk/histmed/ publications/wellcome-witnesses/index.html or by following the links to Publications/Wellcome Witnesses at www.ucl.ac.uk/histmed

A hard copy of volumes 21–32 can be ordered from www.amazon.co.uk; www.amazon.com; and all good booksellers for £6/$10 plus postage, using the ISBN.

Other publications

Technology transfer in Britain: The case of monoclonal antibodies
In: Tansey E M, Catterall P P. (1993) *Contemporary Record* **9:** 409–44.

Monoclonal antibodies: A witness seminar on contemporary medical history
In: Tansey E M, Catterall P P. (1994) *Medical History* **38:** 322–7.

Chronic pulmonary disease in South Wales coalmines: An eye-witness account of the MRC surveys (1937–42)
In: D'Arcy Hart P, edited and annotated by E M Tansey. (1998) *Social History of Medicine* **11:** 459–68.

Ashes to Ashes – The history of smoking and health
In: Lock S P, Reynolds L A, Tansey E M. (eds) (1998) Amsterdam: Rodopi BV, 228pp. ISBN 90420 0396 0 (Hfl 125) (hardback). Reprinted 2003.

Witnessing medical history. An interview with Dr Rosemary Biggs
Professor Christine Lee and Dr Charles Rizza (interviewers). (1998) *Haemophilia* **4:** 769–77.

Witnessing the Witnesses: Pitfalls and potentials of the Witness Seminar in twentieth century medicine
By E M Tansey. In: Doel R, Soderqvist T. (eds) (2006) *Writing Recent Science: The historiography of contemporary science, technology and medicine.* London: Routledge, 260–78.

INTRODUCTION

No therapeutic drug, before or since, has captured the public imagination in the way that penicillin did in the years immediately following the Second World War. This remarkable drug had been brought to the world by the brilliant work of Howard Florey, Ernst Chain, Norman Heatley and their colleagues at the Sir William Dunn School of Pathology in Oxford at the start of the war. Thanks largely to the vital input of American biological, chemical and commercial expertise, sufficient penicillin was produced during the war to influence the treatment of battle wounds (and that other common cause of war casualties, gonorrhoea) and to provide for widespread civilian use once peace had been restored. It was Alexander Fleming who in 1928 had fortuitously discovered the remarkable antibacterial properties of a substance produced by a contaminant *Penicillium* mould, and though he failed to realize the full implications, he received most of the public adulation: Kevin Brown lists 172 honours showered upon him from all over the world between 1943 and his death in 1955.[2]

Penicillin transformed the treatment of many infections. High on the list of notorious microbial felons successfully targeted by the new drug was the golden staphylococcus – *Staphylococcus aureus* – a germ many of us carry harmlessly in our noses, armpits or other moist areas of the body, but which can cause devastating infection when given the opportunity. The vast array of disease caused by this archetypal bacterial pathogen ranges from boils, carbuncles and abscesses, through toxin-mediated conditions such as scalded skin syndrome, toxic shock and food poisoning, to septicaemia, wound infection and osteomyelitis. In the laboratory, the staphylococcus is an unpredictable tease. I can affirm from personal experience that, whereas Gram-negative bacilli like *Escherichia coli* generally respond reliably and reproducibly to exposure to antibiotics in *in vitro* tests, no two strains of staphylococci seem to behave the same, and the responses of individual strains often appear maddeningly irreproducible. It may be no coincidence that Fleming made his celebrated discovery of penicillin while engaged in a frustrating study of the properties of staphylococcal variants.

Fleming, in his acceptance speech for the Nobel Prize in Physiology or Medicine, which he shared with Florey and Chain in 1945, warned that inappropriate use of penicillin might lead to the development of resistance, something that could

---

[2] Brown (2004): 213–17.

be readily generated in laboratory tests.[3] However, when resistance to penicillin among strains of *Staphylococcus aureus* did indeed become a major problem in hospitals in the 1950s, the cause of the resistance was not the mutational or adaptive changes that Fleming had envisaged, but the selection of strains of staphylococci able to produce an enzyme – penicillinase (β-lactamase) – that had been described as early as 1940.[4] It was the prevalence of strains of this type that, in the late 1950s and early 1960s, largely stimulated interest in developing semi-synthetic penicillins (and related cephalosporins) that were able to withstand the enzyme. The first fruit of this research, methicillin, was produced at Beecham Research Laboratories in 1959 and marketed the following year. It was soon followed by more active compounds, such as cloxacillin and its relatives (isoxazolylpenicillins), which had the additional advantage of being absorbed orally. These events are well documented in an earlier Witness Seminar.[5]

As described during the present Witness Seminar, resistance of *Staphylococcus aureus* to methicillin was detected within a year of the antibiotic appearing on the market. This time Fleming's prediction was closer to being realized, since the resistance was not associated with β-lactamase or any other drug-destroying enzyme, but was eventually shown to result from a mutational change in the proteins attacked by penicillin, which allowed the bacteria to avoid the consequences of exposure to the drug. These variant strains became known as 'methicillin-resistant *Staphylococcus aureus*', or 'MRSA'. However, methicillin has long been superseded by the more active derivatives and is no longer marketed. Since these strains nowadays usually exhibit resistance to several other commonly used antibacterial agents, the acronym MRSA would be better employed, as Professor Gordon Stewart notes in his introductory statement, to refer to them as 'multi-resistant *Staphylococcus aureus*'.

Alarmingly, it soon became clear that methicillin resistance extended to all members of the expanding β-lactam antibiotic family, and this has remained true as newer compounds of this type have been developed. At first the resistance was not perceived to be a major problem: it appeared to be uncommon and seemed to require unusual, non-physiological conditions in laboratory tests – reduced incubation temperature or an abnormally high salt concentration – for full expression of the resistance phenotype, leading to dispute about the clinical significance of the phenomenon.

[3] Freely available at http://nobelprize.org/nobel_prizes/medicine/laureates/1945/fleming-lecture.pdf (visited 30 January 2008).

[4] Abraham and Chain (1940).

[5] Tansey and Reynolds (eds) (2000).

However, during the late 1960s and beyond, the true extent of the problem became clear in Britain and many other countries.

MRSA has now been around for close on half a century, but has only lately become a cause of deep public anxiety, especially among those faced with admission to hospital. The concern has been fuelled by frequently hyperbolic reports in the media (sometimes by microbiologists who ought to know better), which have mythologized the organism as a 'superbug', whereas it is a conventional pathogen merely responding to the intense selective pressure of antibiotic usage through a classic Darwinian survival process. This is not to diminish its importance. MRSA – just like its parent, methicillin-sensitive *Staphylococcus aureus* – is a formidable micro-organism capable of deploying a battery of virulence factors to cause serious, even life-threatening, disease. Its special significance lies in the reduced options for treatment open to the prescriber. Many preferred choices are excluded by the resistance profile (and valuable time lost identifying appropriate alternatives) though, contrary to popular belief, there are usually several drugs other than the commonly preferred vancomycin to which the organism remains susceptible.

After nearly 50 years, the time is ripe for recording personal experiences of the many fascinating twists and turns that the story of MRSA has undergone. Not that the story is yet complete, as this Witness Seminar amply demonstrates: the expert participants often found themselves tempted to turn the discussion to current, rather than historical problems. This is no bad thing, since the present will quickly become 'history' and it is as useful to have a record of contemporary as well as past concerns. Moreover, the 'story so far', in all its multi-faceted aspects, clearly emerged from a very lively discussion with the sort of passion that only those who have to live with the problem during their professional lives can provide. That is the great strength of these seminars, which continue to provide an invaluable record of key topics in recent medical history. As always, we are immensely grateful to Tilli Tansey, Lois Reynolds and their colleagues and to Robert Bud who was instrumental in bringing together a representative group of highly articulate experts to record for posterity their thoughts, reminiscences and comments on an important medical event of our times.

David Greenwood
University of Nottingham

SUPERBUGS AND SUPERDRUGS:
A HISTORY OF MRSA

The transcript of a Witness Seminar held by the Wellcome Trust Centre for the History of Medicine at UCL, London, on 11 July 2006

Edited by L A Reynolds and E M Tansey

SUPERBUGS AND SUPERDRUGS: A HISTORY OF MRSA

Participants

Professor Graham Ayliffe
Dr Robert Bud (Chair)
Professor Mark Casewell
Dr Bilwanath Chattopadhyay
Dr Stephanie Dancer
Dr Bernard Dixon
Dr Georgia Duckworth
Professor Brian Duerden
Professor Michael Emmerson
Professor Gary French
Professor Curtis Gemmell
Professor Alan Glynn
Dr Ian Gould
Professor David Greenwood
Professor Jeremy Hamilton-Miller

Dr Angela Kearns
Dr Bill Newsom
Professor Ian Phillips
Dr Tyrone Pitt
Dr Elizabeth Price
Professor Sir Mark Richmond
Dr Geoffrey Scott
Dr Joe Selkon
Dr David Shanson
Dr Norman Simmons
Professor Dale Smith
Professor Brian Spratt
Professor Gordon Stewart
Dr Robert Sutherland
Professor John West

Among those attending the meeting: Dr Helene Joffe, Professor Ian McDonald,[†] Professor Anne Murcott, Dr Brigitte Nerlich, Dr Sujatha Raman, Dr Judith Richardson, Mr Peter Washer

Apologies include: Professor Sir Roy Anderson, Dr F R Batchelor, Mrs Christine Beasley, Professor Sir Christopher Booth, Professor Bill Brumfitt, Professor James Burnie, Dr Rosa Coello, Dr Martin Cole, Professor Mary Cooke, Dr Barry Cookson, Dr Natasha Crowcroft, Professor Naomi Datta, Professor Susanna Eykyn, Professor Leon Fine, Dr Harold Gaya, Professor Harold Lambert, Dr Owen Lidwell, Professor Edward Lowbury,[‡] Professor John Macfarlane, Dr Graham Marshall, Professor Richard Novick, Professor Francis O'Grady, Professor Hugh Pennington, Professor Richard Proctor, Dr Richard B Roberts, Dr George Rolinson, Dr Christian Ruef, Dr David Self, Professor Reginald Shooter, Sir Richard Sykes, Dr Morag Timbury, Mr Rami Tzabar, Dr Diana Walford

[†] Died 13 December 2006

[‡] Died 10 July 2007

Dr Tilli Tansey: Good afternoon, ladies and gentlemen, and welcome to this Witness Seminar on the history of MRSA. Our Chairman today is Robert Bud, who also chaired an earlier meeting on post-penicillin antibiotics.[1] He is a well-known historian of modern medical sciences at the Science Museum, London, and he is about to publish a new book called *Penicillin: Triumph and tragedy*, which I notice today is already on the Amazon website, to be published in January 2007.[2] So without further ado, I will hand over to Robert.

Dr Robert Bud: Thank you very much, Tilli. It is a great pleasure to welcome so many people who have been my mentors and advisers, and people whose work I have read and haven't yet had the chance to meet. I think the key thing today is not to feel that you have to be called upon, but rather, spontaneously be moved to speak. I am familiar with the possible contributions of a few here, so please do contribute.

We have a series of subject-based headings [in the programme] and I suspect we will move rapidly between periods.[3] Don't be embarrassed about that. If you feel that you have a contribution about the 1960s or about the 1990s, then don't feel that if the conversation seems to be about the 1960s, that you are not allowed to talk about the 1990s. On the other hand, I think that we do want to keep the conversation broadly historical, so while the contemporary issues, particularly the opportunities and challenges of the year 2006, will be important, and a part of the historical debate, I think it will be a shame if these dominate, because there are so many other venues where particular contemporary issues are being debated.

In a sense, my contribution to this was above all to discover Gordon Stewart, author of the book, *The Penicillin Group of Drugs*, which was published in 1965, where the story of the discovery of methicillin-resistant *Staphylococcus aureus* (MRSA) was first outlined.[4] I think we are very privileged that Gordon is willing to kick off and give a brief introduction.

---

[1] See Tansey and Reynolds (eds) (2000).

[2] Bud (2007).

[3] Witness Seminar Programme, 11 July 2006: Historical Introduction; Origins and Early History: Discovery/encounter with MRSA; Mode of transmission; Scottish perspective; Geography of spread; Epidemiology and surveillance; Science: Differentiation of pathogenic and carrier strains; Mechanisms of resistance; Genetics, organism and host; Pharmaceutical response; Surgery: Changing attitudes to infection; Epidemiology: Distribution and clinical pattern of infections; Staph. Reference Laboratories; Hospital response: Nursing; Problems of control: chemotherapy; new drugs; Public response/1990s.

[4] Stewart (1965).

Professor Gordon Stewart: Thank you, Robert, I am very glad to have been discovered; I thought that I was obsolete. That could still be true, as will be apparent in our discussions.

But I would prefer to remind you that we are here mainly because of non-obsolescent discoveries, like the new biosynthetic penicillins developed in 1958 and 1959 by the Beecham group of research workers.[5] We are fortunate in having Dr Sutherland from that talented team here with us today. Unfortunately, others have died since the last seminar in 1998.[6] There are many others who were invited, but for other reasons cannot be present. We certainly must pay tribute to their memory, because without them and all those drugs, especially methicillin, there would be no methicillin-resistant *Staph. aureus* (MRSA), now promoted to international status as the notorious multidrug-resistant *Staph. aureus* (also MRSA) and epidemic MRSA (EMRSA). Methicillin itself is also obsolete, except as an infallible laboratory identifying marker and label for this unusual and terrifying organism.

Figure 1: Dr Frank Batchelor and Dr Peter Doyle, two of the original members of the Beecham group of research workers, at the 1998 Witness Seminar, 'Post Penicillin Antibiotics'.

---

[5] See Tansey and Reynolds (eds) (2000): 5, 28, 31, 37, 63.

[6] Peter Doyle (1921–2004), David M Brown, John Nayler (1927–93) and Eric Knudsen (1921–2001) of Beecham; Joseph Anthony Porteus Trafford (Baron Trafford of Falmer from 1987) (1932–89) and Arthur Henry Douthwaite (1896–1974) at Guy's; and my colleague Richard J Holt (1920–2005) at Queen Mary's Hospital. See biographical notes. See also Rolinson (1998).

It wasn't terrifying to begin with. In 1961 when the first strains were isolated, they didn't seem to matter very much.[7] Essentially, they were commensal inhabitants of nostrils and skin flexures, which spread without doing much harm – until an outbreak occurred and spread with amazing rapidity within one hospital.[8] There were two hospitals in London using methicillin in 1959/60, Guy's and Queen Mary's, Carshalton, where about 40 children became infected quickly.[9] Even so, it was mainly a commensal and superficial infection, until we had one fatal case, who died with fulminating septicaemia, confirmed at post mortem as a group III strain of MRSA.

Fortunately, we put this strain on file in 1961, followed by a second from an outbreak in 1962 with another death. This put a different complexion on it, because other people reported similar incidents and put similar strains on file and attracted attention under various headings. It wasn't just the fact that it caused a few severe cross-infections. It was also the fact that it was a very new kind of organism and many of you who are here today must know much more than I do about this, especially Graham Ayliffe, who will be able to tell you more about it.[10] But what we did find out then was that this was an organism, which in colonial characteristics, in molecular structure and in other ways, appeared to be very different – Mark Richmond might speak about that. It had a different kind of structure altogether. It wasn't like the ordinary *Staph. aureus*, it seemed to be primitive.

During the first outbreak, we found that it could revert very quickly to a primitive form, rather like L-forms which could become syncytial without a cell wall as a boundary. There was no way of saying this could be linked to RNA and chromosomes, it grew all over the plates, and in fact Mark Richmond commented later that it didn't seem to be chromosomal. He didn't at that time, I think, say for sure that it was episomal or extrachromosomal. However, the Americans and the Japanese – especially Keiichi Hiramatsu in Tokyo and Richard Roberts at Cornell, Richard Proctor in Wisconsin and a few others – found later in the 1990s that there was a chromosomal factor after all.[11] DNA was in the picture, as it is in most things. That factor, the *mecA* and allied genes,

---

[7] Jevons (1961).

[8] Stewart and Holt (1963).

[9] Knox (1960); Stewart *et al.* (1960).

[10] See page 7.

[11] Hanifah and Hiramatsu (1994); Roberts *et al.*, Tri-State MRSA Collaborative Study Group (2000).

is now known to be responsible for the virulence, especially in septicaemic and surgical infections, because these are the biggest problems.[12] There's no point going into the detail of all that now, although the detail is fascinating and we will hear more about it later.

In other ways, the challenge is largely ecological. The staphylococcus is an inhabitant of skin flexures and found in different situations, in hospitals, especially in theatres, and often in the noses of doctors, nurses and others. It took a long time before people would admit that this was something that should be dealt with. S D Elek and others at St George's Hospital Medical School, London, were very quick on the draw there, and they tried to devise ways of preventing the spread and, to some extent, were successful.[13] But MRSA continued to spread in a series of small outbreaks, firstly in Europe and then in South Africa, Egypt and a few other countries. And then a strange thing happened. In 1964, Jerzy Borowski in Bialystok, eastern Poland, found that this was a strain that had existed before any methicillin was used.[14] We had been using methicillin, but he hadn't, nor had some of the Russians nearby, although shortly after that they got hold of some, goodness knows how. Then we found that the same kind of resistance, complete with fatal cases, was occurring in situations with methicillin and the newer penicillins, like the isoxazolyls and so on, that had *not* been in use. So this was a new clue that's been followed through and shows that we are now dealing with something that is chromosomal, but also environmental and is extra chromosomal, but not closely related to therapy, as further forms of resistance are.[15] The overseeing of multi-resistance is hard enough, and is very widespread, but fortunately, we have vancomycin, complete with a new acronym, VRSA (vancomycin-resistant *Staphylococcus aureus*), as well. The only drugs that seem to be controlling this – and we knew this back

---

[12] Hurlimann-Dalel *et al.* (1992).

[13] See, for example, Stern and Elek (1955); Elek and Conen (1957); Elek and Fleming (1960). Elek also contributed the Elek-Ouchterlony gel diffusion method for detection of anti-alpha haemolysin and anti-leucocidin. Elek (1948a and b); Ouchterlony (1948).

[14] Borowski *et al.* (1964, 1967); Borowski (1988a and b).

[15] Professor Gordon Stewart wrote: 'After publication of the success of methicillin in 1960, it was used more extensively and more freely in major centres in the US than anywhere else through the 1970s. Tens of thousands of strains of *Staph. aureus* were tested, but there were very few isolations of MRSA, no reports of indigenous outbreaks until the 1980s and no major incidents until the 1990s when strains with the *mecA* gene and other markers of virulence were identified in strains isolated from severe and fatal cases.' Note on draft transcript, 8 December 2007.

in 1960 – were the quinoline-related amides, such as quinacillin, and, more recently, daptomycin.[16] Professor Emmerson might say something about that. There are other drugs in use, including, oddly enough, some new twisted structures linking a quinoxolidone structure to penicillins and small peptides.[17]

Now, both of these have turned up historically, but there were many other things that we had in mind when designing this Witness Seminar, a follow-up to the earlier meeting in 1998, also supported by the Wellcome Trust, when MRSA was not such a problem, although spreading rapidly.[18] The meeting today is a sequel to that, which Robert Bud steered so well then and now. You are the witnesses to speak about these different aspects. We are hoping to cover all the main topics, and that's saying something, including ecological, genetic, microbiological, and administrative issues – not least administrative, because, to a large extent, the problem is one of full hospitals, over-full bed lists, transfer of cases between wards and this kind of thing. The human story is made very sad by all the incidents that are linked to these.

Bud: Thank you very much, Gordon. I think we have some people here who have been engaged with MRSA since the very first experience of it. Graham Ayliffe has very kindly agreed to kick off for us, while people are thinking about what they can say.

Professor Graham Ayliffe: We ought to start by saying a few words about the situation before methicillin appeared. Superbugs have long been a cause of hospital infection. Hospital gangrene, pyaemia and erysipelas caused by haemolytic streptococci and *Staph. aureus* were responsible for many hospital outbreaks in the nineteenth century and had a high mortality. When penicillin came along we thought this was the end of staphylococcus as an important cause of hospital infection. But, as you all know, resistant strains soon appeared and Mary Barber described them at the Hammersmith Hospital, London, in the 1940s.[19] There were other reports in the US, and particularly in Australia.[20]

---

[16] See, for example, Sato *et al.* (1967); Gedney and Lacey (1982); Maple *et al.* (1991); Sheldrick *et al.* (1995).

[17] Wasserman (2006).

[18] See Tansey and Reynolds (eds) (2000).

[19] Mary Barber found that 38 per cent of strains of *Staphylococcus aureus* were penicillin-resistant in 1947, increasing the following year. See Barber (1947).

[20] See Rountree and Barbour (1950, 1951).

What was very important at this time was the introduction of staphylococcal 'phage-typing, which enabled us to investigate these outbreaks.[21] And then the particular virulent strain of 80/81 appeared, I think, first of all in Australia, described by Phyllis Rountree.[22] This strain caused a lot of superficial and severe sepsis, mainly in maternity units, causing severe sepsis in babies, furunculosis and carbuncles in staff and abscesses in nursing mothers. Then, this tended to disappear over the 1950s.

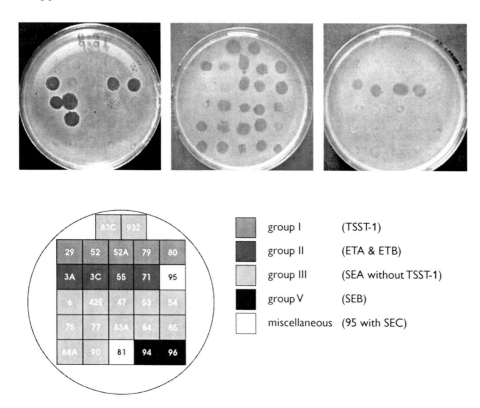

Figure 2: Bacteriophage-typing, c. 2003: three Petri dishes showing different strains of S. aureus tested at 100 × Routine Test Dilution (RTD), giving varying patterns of lysis with 27 'phages (including 23 'phages of the International Typing Set and 4 experimental 'phages), as shown in the reference grid. From slides provided by Dr Angela Kearns.

---

[21] Blair and Williams (1961); Hartstein et al. (1995). See also Cox et al. (1995). See Glossary, page 125; see also Figure 2.

[22] Rountree and Freeman (1955). See Appendix 2, page 82.

Figure 3: Directors of the PHLS at R E O Williams' retirement in 1985.
Seated: Sir Graham Wilson (1941–63); standing L to R: Dr R E O Williams (Staphylococcus and Streptococcus Reference Laboratory, 1949–60; PHLS, 1973–81); Dr J E M Whitehead (1981–85); Dr Joe Smith (1985–92).

But in surgical wards there were other problem epidemic strains. These were 'phage group III strains, not so virulent as the 80/81, and really were the start of most of our present problems with resistant strains.[23] They were virtually all resistant to tetracycline, but there were strains acquiring increasing resistance to erythromycin, novobiocin, neomycin, etc.[24] We ended up having some strains which were resistant to most of the available antibiotics.

Before going on, I should mention the large amount of work done in the 1940s to the 1960s, which formed the basis of all our knowledge of staphylococcal cross-infection and prevention. Among these, Robert Williams, Reggie Shooter and Robert Blowers were prominent and most of this story is summarized in a comprehensive book.[25] There were others, such as Owen Lidwell, Edward

---

[23] Professor Graham Ayliffe wrote: 'Staphylococcal bacteriophages are viruses which lyse [destroy] specific strains of *Staphylococcus aureus*. A set of different bacteriophages [see Appendix 2] is applied to strains of staphylococci and varying patterns of lysis are obtained with different strains, e.g. 80/81, 6/7/53/75/77.' E-mail to Mrs Lois Reynolds, 3 December 2007. See Shooter *et al.* (1958); Williams (1959); Parker and Jevons (1963); Parker *et al.* (1974).

[24] Shanson (1981).

[25] Williams *et al.* (1960, 1966).

Lowbury, Tom Parker and Bill Noble, who were very much involved in these early days in describing the carriage sites (noses, perineum) and the heavy dispersers of staphylococci; all these studies were done before the emergence of MRSA.[26]

Initially, I was involved at Bristol in the mid-1950s with William Gillespie on the control of wound outbreaks and although isolation was well accepted at this stage, we were concerned with preventing infection in wards where no isolation was possible.[27] Topical antiseptics and antibiotics, such as neomycin and chlorhexidine, were applied to noses and wounds, plus some environmental improvements, such as sterilization of blankets, etc. We found in those days that a single measure by itself was very unlikely to be successful; also if you removed an infected patient from the open ward, the outbreak tended to go away. But by 1960, although we had some measure of control, this was a period of disenchantment with antibiotics, and in those days only these highly toxic antibiotics (vancomycin and ristocetin) were available to treat these highly-resistant strains. With the introduction of methicillin in 1960 – marketed as 'Celbenin' – many believed that this was the end of the resistant staphylococcus and Ernst Chain was the one who said 'no more resistance problems, methicillin is the answer'.[28] Mary Barber was one of those who disagreed with him, quite rightly, that this wasn't the answer to the problem.[29] Patricia Jevons, working at the Central Public Health Laboratory in Colindale, as we have already

---

[26] For further details, see Tansey and Reynolds (eds) (2000).

[27] Professor Graham Ayliffe wrote: 'In the Bristol Royal Infirmary [Gillespie *et al.* (1961)] at that time, noses and the perineum were recognized as important carriage sites, but major sources were usually the dangerous spreaders, who became known as dispersers. These dispersers were associated with heavily discharging wounds, pneumonia or generalized dermatitis and were identified by Ronald Hare and others [Hare and Thomas (1956)]. Bill Noble and R Davies made the important discovery that staphylococci were spread on skin scales [Davies and Noble (1962)]. We often found that spread could occur from hidden sources such as bedsores and outbreaks usually ceased when a disperser was removed from the ward.' Note on draft transcript, 9 January 2007.

[28] Ernst Chain quote, personal communication to Mary Barber, recalled by Professor Graham Ayliffe. See also Knudsen and Rolinson (1960); Rolinson *et al.* (1960).

[29] Dr Mary Barber was influential in formulating an antibiotic policy for the Hammersmith Hospital in 1958, along with representatives of a Joint Committee of the Medical Research Council Antibiotics Clinical Trials Committee and the staff of the Hammersmith Hospital: Professor L P Garrod (Chair), A A C Dutton, P C Elmes, W Hayes, P Hugh-Jones, E J L Lowbury, J G Scadding, R Schackman and J P M Tizard. For their regime, see Barber *et al.* (1960): 12. Other guidelines on antibiotic use have been published by committees of the Hospital Infection Society, British Society for Antimicrobial Chemotherapy and the Infection Control Nurses Association in 1986, 1990, 1998 and 2006.

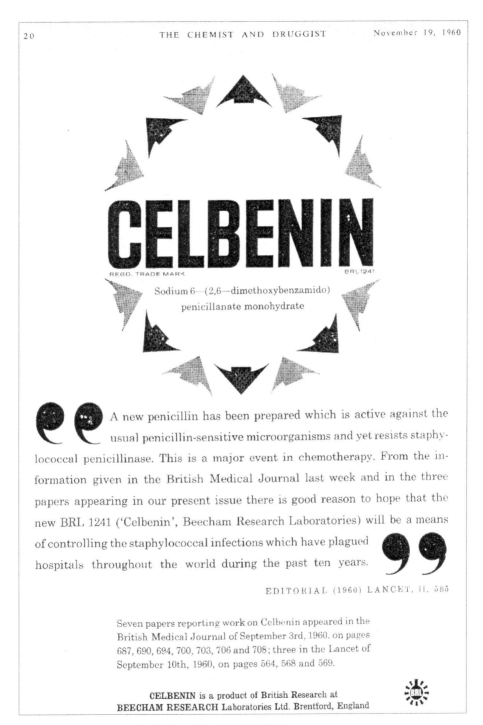

Figure 4: Advertisement for 'Celbenin' (methicillin), 1960.

mentioned, looked at 5000 strains, and found a methicillin-resistant strain in one hospital.[30] I think this didn't worry people too much, because it was really still quite a rare event.

At that time, many of the investigations were done by Mary Barber. I was working with her from about 1960 and we were joined by Pamela Waterworth after Professor Garrod retired.[31] Mary Barber studied a number of strains of naturally occurring MRSA, mainly obtained in this country, but also from Yves Chabbert in Paris and Kirsten Rosendal in Denmark.[32] She found that they were most closely related to 'phage group III strains from all these countries. They were all resistant to penicillin and tetracycline, but they had an unusual heterogeneous resistance. In the presence of low concentrations of methicillin, a heavy growth was obtained where there was a heavy inoculum on the plate, whereas individual colonies were inhibited on other parts of the medium. Mary Barber showed that the addition of an electrolyte of sodium chloride – 5 per cent – actually enabled these colonies to grow normally. It seemed likely that methicillin interfered with cell wall synthesis. D I Annear in Australia showed that growing these MRSA at 30°C would also provide a normal growth.[33] We, and others, also found that these strains produced high levels of penicillinase, but methicillin was really quite resistant to penicillinase. The methicillin-resistant strains produced in the laboratory were different, and these naturally occurring strains were virulent, both in animals, and, as Professor Stewart said, in surgical patients.

Both Jerzy Borowski from Poland and Ozdem Äng from Turkey were working at the Hammersmith Hospital, London, at the time, and they returned home and started looking for MRSA. As Professor Stewart said, they both found them, and MRSA was also found in India this time.[34] So this indicated that methicillin resistance wasn't related necessarily to the clinical use of methicillin. These strains were also multi-resistant, which was also something that needed an explanation, although this further indicated that MRSA was in existence before methicillin was introduced, but did not explain the multi-resistance of these early strains. Clinical infections were increasingly reported during the 1960s, usually in large hospitals, mainly in Europe, Denmark, Switzerland, but strangely enough at

---

[30] Jevons (1961).

[31] Barber (1961).

[32] Barber (1964a).

[33] Annear (1968).

[34] Pal and Ray (1964).

that time, not so often in the US.[35] Other penicillinase-resistant penicillins, and the cephalosporins, were introduced during that decade. Unfortunately, Mary Barber died in a road accident in 1965 and this was at an early stage of her work on this subject.

Bud: Is there somebody else? I know that Professor Hamilton-Miller was involved very early on.

Professor Jeremy Hamilton-Miller: I was at Guy's in 1960 and, as Professor Stewart has just said, we were doing one of the first clinical trials there, and he was doing his at Carshalton [Queen Mary's Hospital for Children].[36] Professor Robert Knox was a consultant to Beecham's and so we got some of the early specimens. I can recall the excitement of the clinical trial that was being done on methicillin by Arthur Douthwaite, Anthony Trafford and others.[37] They had this magic white powder, we didn't know its structure and we weren't told what its pharmacokinetics were, we were just given the phials and told to inject it every six hours, a far cry from modern clinical trials, I can assure you.[38] But

---

[35] Professor Graham Ayliffe wrote: 'Clinical infections were increasingly reported during the 1960s. A prevalence survey carried out by the Hospital Infection Research Laboratory [Ayliffe (1973)] of noses of patients in general hospitals of the West Midlands showed an increase from 0.14 per cent of 1478 patients in 1967 to 2.7 per cent of 2321 patients in 1970. Tom Parker from Colindale at the first major International Conference on Hospital Infection in Atlanta in 1970 [Parker (1971)] reported that 5 per cent of staphylococcal isolates in Britain were resistant and Kirsten Rosendal in Denmark reported that 40 per cent of staphylococcal isolates from blood cultures were MRSA [Rosendal (1971)]. The increase in methicillin resistance was not related to an increase in usage of penase-resistant penicillins, which were introduced during this decade, but all showed cross-resistance with methicillin.' Note on draft transcript, 4 December 2007.

[36] Stewart *et al.* (1960).

[37] Douthwaite and Trafford (1960). See Figure 4.

[38] Professor Jeremy Hamilton-Miller wrote: 'Here are some parameters, if you wish to include them: half-life: 0.5–1 h[our]; peak level after 1g IM [intramuscular injection] $\geq$ 10mg/L[itre]; 40 per cent serum bound; C > MIC [simulated peak plasma concentration (C) to the literature value of the minimum inhibitory concentration (MIC)] for 3–4 hours following 1g IM.' Note on draft transcript, 1 December 2007. Professor Gordon Stewart wrote: 'I worked closely with Dr John Farquharson, an organic chemist who was the first Director of Research at Beecham's. We knew that we were handling derivatives of 6-APA with insertions of amino- and methyl-groups which altered antimicrobial activity. The formula for BRL1241, methicillin, was divulged to me in confidence after we had used it to arrest an MRSA septicaemia in December 1959. The difference between the D(-) and L(+) epimers of BRL1060, ampicillin, was recognized by myself before the therapeutic trial. Although local or individual circumstances may sometimes justify exceptions, safety in trials should take precedence over commercial secrecy and patents.' Note on draft transcript, 8 December 2007. See also Stewart (1965): 120; note 144.

the surgeons were absolutely delighted, because patients had been dying of staphylococcal sepsis at the time.

I believe the first MRSA isolated was from Guildford in 1959. It was given the number 13136 by the people at Colindale and there were two letters published in the *British Medical Journal*, the first by Jevons, and the second by Knox.[39] The one by Jevons is always quoted, although they were actually side by side, because 'J' comes before 'K' in the alphabet, which I always thought was a bit unfair on Robert Knox. We didn't find any MRSA at Guy's, unlike Stewart's team at Carshalton.[40] I sometimes wonder about this: looking back at those days, we didn't actually know that you had to reduce the temperature to 30°C or add salt. At that time, as Graham [Ayliffe] pointed out, that wasn't actually made clear until Bob Sutherland and Mary Barber showed these requirements.[41] I wonder if those figures that Patricia Jevons came out with – that 0.05 per cent of strains that they looked at retrospectively were methicillin-resistant – might actually have been a bit higher. You certainly do miss them if you grow them at 37°C under normal laboratory conditions. There was a tremendous controversy as to whether these MRSAs were hyperproducers of β-lactamase and Knox and Smith did quite a lot of work on that.[42] I think we all came to the conclusion, as did Mark Richmond and his team, that this wasn't the case. Although methicillin in high concentration is labile, it is not at the concentrations attained therapeutically.[43]

Because of the heterogeneous nature of MRSA, there was considerable discussion as to whether you could use high doses of the cephalosporins. At one stage there was quite a lot of support for using cefamandole, the cephalosporin which is most active against *Staphylococcus aureus*.[44] As Professor Stewart has said, the problem virtually disappeared, until we got to the EMRSA (epidemic MRSA) story, which is a bit different.

Professor Michael Emmerson: I would like to add to Graham Ayliffe's and Gordon Stewart's points. There was a bit of a smokescreen at that time, because

---

[39] Anon. (1961); Jevons (1961); Knox (1961); Rolinson (1998).

[40] Stewart *et al.* (1960).

[41] Sutherland and Rolinson (1964); Barber (1964). See also page 12.

[42] Knox and Smith (1963).

[43] Hamilton-Miller and Ramsay (1967).

[44] Eykyn *et al.* (1973).

we were still coping with 'phage 80/81 and we wanted to know what the virulence or pathogenic markers were. Phyllis Rountree was in London at the time and was looking at cadmium and mercury resistance. I was a senior house officer at University College Hospital (UCH), London, with David Shanson, and Dr Joan Stokes employed us to take swabs from members of staff who came into the hospital, so we were surrounded by large numbers of nasal swabs, looking for tetracycline resistance. Like Jeremy [Hamilton-Miller], we didn't know about the effects of temperature and salt etc., so we were looking for cadmium and mercury markers and the carriage rate among medical students, nurses and the like. We weren't concerned with methicillin, because the numbers were so small at the time, we still wanted to know why 'phage 80/81 killed young people, particularly young men going for clean operations without risk factors. And then, of course, some of these were tetracycline-resistant and not penicillin-resistant, and so we were still very confused at that time. I think that may be why our eyes were off the ball, so to speak, round about the mid-1960s.

Professor Ian Phillips: I was appointed as an assistant lecturer in the Department of Microbiology at St Thomas' in 1963, and one of the first jobs that I was given was to look at stored clinical isolates of staphylococci from within the hospital, by the method that has been mentioned by Graham [Ayliffe], Mary Barber's salt plates.[45] Eventually I trawled 15 isolates between 1961, when our collection started, and 1963 when I did the investigation. The interesting thing is that we had missed all of them in our ordinary susceptibility testing, which again prompts me to wonder whether, say, the original Beecham screens on staphylococci had missed isolates as did Patricia Jevons'. So, we had missed them in the laboratory. But interestingly, we had also missed them clinically, so there was never any question from the clinicians that what was being used was failing. Looking back, they were proper infections. Indeed, I remember one patient who had an osteomyelitis, which surprised me. I am not quite sure what the epidemiology of that one could have been, but that's what it was. So, yes, we missed them and I wonder how far back they really go, apropos of what Gordon was saying. Were they selected by methicillin, or were there strains before?[46]

---

[45] Barber (1964b).

[46] Professor Gordon Stewart wrote: 'Studies in the UK, US and Poland suggested that strains with natural resistance existed before methicillin was used in volunteers or clinically. Surprisingly, all of them formed constitutive penicillinase, were resistant also to tetracyclines and showed differences in cellular and colonial morphology.' Note on draft transcript, 8 December 2007. See comments by Duckworth (page 22), Phillips (page 24) and Richmond (page 35).

Dr Joe Selkon: I started work at the Hammersmith Hospital [London] in 1954. After initially working on *Mycobacterium tuberculosis*, including a four-year spell in India, I thought I had better learn about general microbiology and so I attached myself to Mary Barber's tuition on the control of hospital-acquired infection. This was a most important education, for she taught us a fundamental lesson that went back to Florence Nightingale: the first requirement of a hospital is that it does no harm.[47] I then went to work in the General Hospital, Newcastle upon Tyne, in 1963 and was amazed at the number of staphylococcal infections occurring in our surgical patients and the complete absence of adequate extract-ventilated rooms for the isolation of patients who had multi-drug resistant staphylococcal infections.[48] I therefore spent three or four years building isolation rooms attached to the wards, with extract ventilation as suggested by the work of Williams and Shooter.[49] However, by 1968 we were in serious trouble, for in that year we isolated staphylococci resistant to methicillin from 37 different patients. Despite everything we did in following the standard requirements as suggested by Williams and Shooter, the situation deteriorated further and MRSA were increasingly isolated from surgical wounds – from 37 in 1968; 104 in 1969; 127 in 1970; 177 in 1972; to 134 in 1973. The bacteriophage type 75/80 predominated, being present in 84 per cent of the 177 MRSA strains isolated in 1972.[50] Quite clearly, our hospital did not meet the requirements of a place that does no harm. The health authority understood our predicament and agreed to build a separate isolation unit of 12 beds. We started studying the epidemiology of methicillin-resistant staphylococcal strains throughout the hospital and found, surprisingly, that MRSA were circulating within the medical wards where no infections were occurring. However, when those patients were transferred into surgical wards, they went down with an MRSA infection after surgery and started an epidemic there. We therefore initially used our isolation ward to clear the presence of MRSA from our medical wards and screened new patients for surgery. Slowly, over two years, we succeeded in reducing the isolation of MRSA from all patients in this 1000-bed hospital, and over the following five years to 14 (0.53%) of the 26 586 admissions.[51] In the

---

[47] Nightingale (1863): Preface. See also Reynolds and Tansey (eds) (2007b):196.

[48] See Kinmonth *et al.* (1958). For details of the clean air theatre introduced for orthopaedic surgery in 1960, see Reynolds and Tansey (eds) (2007a).

[49] Shooter *et al.* (1958).

[50] Selkon *et al.* (1980).

[51] Ingham (2004).

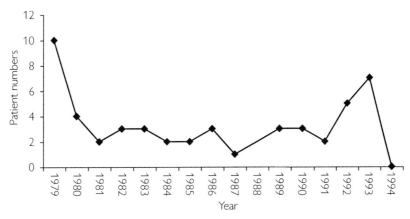

Figure 5: Number of patients with MRSA admitted annually to the isolation ward at Newcastle General Hospital, 1979–94. Adapted from Ingham (2004).

ten subsequent years to 1994 there were seldom more than three patients with MRSA infections in any one year [see Figure 5].

Unfortunately, in 1995 the Newcastle upon Tyne health authority [Trust for the Newcastle General Hospital] decided to use the isolation wards to treat AIDS patients only, since they believed MRSA were 'no longer a problem'.[52] Within a number of years the MRSA figures were back to those we started off with. I mention this as an historical picture of what happened when we used the appropriate facilities and the failure to adhere to these fundamental principles of isolation by the administration of the health service after 1994.[53]

Professor Sir Mark Richmond: One or two small points: I got involved when I returned from Denmark on 1 January 1960 to be confronted with the arrival of methicillin. I was interested in penicillinase and staphylococcal resistance. One person who hasn't been mentioned, who I think was tremendously important but very much behind the scenes, was Tom Parker and his colleagues at Colindale [PHLS's Staphylococcal and Streptococcal Reference Laboratory]. For a while, he and I spent quite a long time trying to work out whether methicillin resistance was actually 'phage-affected. Because, as you know, the typing patterns in staphylococci reflect, to a certain extent, the sensitivity to 'phages, but more often

---

[52] Professor Ian Phillips wrote: 'It is worth noting that MRSA remained uncommon in many hospitals despite a lack of isolation facilities and the application only of routine infection-control procedures, until the advent of EMRSA. At that stage, incidence jumped from 0.2 per cent to 11 per cent of hospital isolates in St Thomas' Hospital. See Phillips (1991).' Note on draft transcript, 6 January 2007.

[53] Selkon *et al.* (1980); Ingham (2004).

Figure 6: Dr Tom Parker, 1982.
Director of the Cross Infection Reference Laboratory, PHLS, 1960–78.

than not it's the resistance patterns from 'phage resistance. It all came to nothing, and nothing was ever published. But Parker, Pat Jevons and Liz Asheshov, who were there at the time, did a tremendous amount of work on that aspect of it.[54]

One other point arising from what Graham Ayliffe said. When I was in Bristol, William Gillespie was the Professor of Clinical Bacteriology there, who had a lot of experience of treating bacterial infections both in the war and after. He always used to rib me, because I was so interested in resistance due to penicillinase: 'Oh well, if you want to get rid of staphylococcal infections, the way you do it is to arrange your patients in the wards'. He saw then that the environment in which the patient was located was a very clear way for dealing with staphylococcal infections. Antibiotics could be useful, but the main thing was to give the patients space: isolate them and have sufficient space between the beds.

Dr David Shanson: In Europe and many parts of the UK in the 1970s there was generally a decline in multiple antibiotic-resistant *Staph. aureus* problems, and people weren't quite sure why this was. Some people postulated that the decreasing use of tetracycline was possibly one aspect. I used to call the 1970s the decade of 'complacency', as far as *Staph. aureus* problems were concerned and the vast majority of sporadic *Staph. aureus* strains were sensitive to methicillin during that decade.[55]

---

[54] See also note 21.

[55] Shanson (1981): 19–21.

The other point about that decade [1970s] is that enormous attention was given to Gram-negative gentamicin-resistant bacteria causing outbreaks of hospital infection. I think the eye was off the ball, in terms of multiple-resistant *Staph. aureus*, until 1976, when there were hospital outbreaks of gentamicin-resistant *Staph. aureus* for the first time all over the world, including Bristol where David Speller reported an outbreak.[56] We at the London Hospital Group reported for the first time an outbreak of *Staph. aureus* due to a methicillin- and gentamicin-resistant strain which was also multiple resistant.[57] We managed to control that outbreak with a package of measures that many people today would regard as way over the top, including the closure of wards, formaldehyde fumigation of closed wards and intensive care units, and the use of a temporary isolation ward to cohort the patients, since there were no isolation facilities. Dr Richard Marples, who was at the PHLS Colindale, mentioned that our particular 'phage type *Staph. aureus* outbreak strain was never seen again, so I think we did a reasonable job.[58]

Dr Norman Simmons: I would like to view the situation with the eyes of a young man in 1961. I was working at Edgware General Hospital with Bill Brumfitt. At the time there was no MRSA. All I can remember is a hell of a lot of patients getting infected. The PHLS 'Survey of Infection' in 1960 showed a 10 per cent post-operative infection rate, and if they had an infection with an organism sensitive to an antibiotic, you gave it to them and they got better.[59] But there were a lot that had an infection with an organism that wasn't sensitive to an antibiotic. They were getting resistance to tetracycline, as well as penicillin, and then the patient got worse and sometimes they died. It's interesting to see that by 1980 the infection rate was said to be 18 per cent, but these comparisons aren't fair, because the patients had different operations. Now, because I don't see Bill Brumfitt here, I want to quote something that he told me that Fleming said to him: 'You know, the staphylococcus is a very clever organism. No matter what antibiotic we find, it will get resistant to it'.

Now what did I think of the organism, staphylococcus, and the antibiotic [methicillin] at that time? I thought simply: 'Here's an antibiotic, we can treat the patients with it, and they will get better'. There was a belief that in becoming resistant to methicillin, the organism somehow lost its pathogenicity.

---

[56] Speller *et al.* (1976).

[57] Shanson *et al.* (1976).

[58] Dr R Marples, personal communication to Dr David Shanson.

[59] PHLS (1960).

That was the hope and I did exactly the same sort of work that Mary Barber did.[60] Unfortunately, she published it just about a week before we finished. You remember that horrible feeling when you have done two years' work and their stuff appears in print before yours. We made organisms resistant to methicillin; they were completely unstable. We got the natural ones and the resistance was stable, and the stable ones were pathogenic in animals. MRSA then became the name of one sort of infection, although there are several sorts of MRSA. Then it became an industry. At first there was the mythology, and now it's become an industry, as if somehow these organisms aren't staphylococci at all. People talk about them as if they don't spread in the same way as other staphylococci; as if they don't do the same things, and yet they do all the things that happened before methicillin appeared, probably before 1960.

If you look at the infection rate in the days of the 1930s and 1940s, it was the same. There were many infections. The most important effect of methicillin was that it put an end to the tremendous amount of research and enthusiasm for infection-control measures which were vigorous and effective at the time, because when people were infected with this nasty organism, you could treat them. Who the hell wanted to put them in an isolation facility? People stopped washing their hands, and doing all the things that they have realized they should be doing now.

Dr Ian Gould: Forgive me, I am too young to reminisce, but perhaps there are one or two contemporary issues which certainly run true with me. Joe Selkon and Mike Emmerson mentioned two points which I might comment on. Joe was talking about how hospitals should do no harm. I am reminded of something I read recently.[61] Of course, coming from Aberdeen I know a bit about Alexander Ogston, the Regius Professor of Surgery there, who discovered the clinical significance of *Staphylococcus aureus* in the late 1870s.[62] I think it is well documented that at the entrance to his wards in the hospital there were signs saying 'Prepare to meet thy maker', or something to that effect. Soon after he discovered *Staphylococcus aureus*, he got involved in Listerian antisepsis and he was soon able to take down those signs from the ward.[63] There's certainly a contemporary ring about that these days when we look at the public anxiety, if not to some extent the reality of the very virulent clones of MRSA that we see, not only in our hospitals but in the community.

---

[60] Barber (1961).

[61] Bulloch (1929).

[62] Ogston (1881).

[63] Lyell (1989).

That brings me to Mike Emmerson's point about why 'phage type 80/81 was killing young people and causing some very serious infections. I think it has been quite well established now that that strain still exists in the form of a so-called sequence-type 30 Asian-Pacific MRSA clone, a close relative of 'phage type 80/81. It has at least one toxin, which probably is also shared with 80/81, called the Panton-Valentine leukocidin, discovered at the London Hospital, I believe, in the mid-1930s.[64] There are two aspects where things have come full circle and are very relevant today, as they were earlier.

Dr Bill Newsom: I have four quick points: Patricia Jevons in her paper, you might remember, said the patients who grew the staph didn't actually have any methicillin, and they had only ever used methicillin in the hospital once on a completely unrelated person.[65] The second is pathogenicity. I will remind you that the first recorded US case was a lab technician who jabbed his finger with a strain that had been sent to them in 1967 from France to be examined, and he had required vancomycin, so he was poorly. The first time we saw it in Papworth Hospital was soon after we started testing for it, and certainly we had one patient who transferred it to three more patients. The first patient came from Ipswich. We rang up Ipswich and they said, 'Ah, I think we are just starting these tests'. They rang back and said, 'Well, actually we have had an outbreak in the orthopaedic ward for some time.' They clearly hadn't realized they had got it. Finally, it was the last speaker's dad [J Cameron Gould, a microbiologist in Edinburgh], who found penicillin in the ward air in Edinburgh.[66]

Professor Mark Casewell: I want to go back to what Graham [Ayliffe] was saying about the studies done in the late 1950s–early 1960s by the Williams–

---

[64] Panton and Valentine (1932). See Glossary, page 128–9.

[65] Jevons (1961).

[66] Dr Ian Gould wrote: 'My father, J C Gould, found that the majority of patients and staff of a large general hospital who were carriers of *Staph. aureus* were found to be colonized with penicillin-resistant strains and non-treated patients also became rapidly colonized with these organisms. This suggested selective pressure in the hospital environment. The report showed that penicillin was present in the environment of the hospital and in highest concentration where it was being used. Droplet nuclei and agitated dust particles containing penicillin were found in the air and penicillin was recovered from fomites and noses of patients and staff in these areas. Examination of a factory handling penicillin showed a similar contamination of the environment with the antibiotic and carriage of penicillin-resistant strains amongst staff. It was concluded that environmental penicillin is an important selective agent leading to the colonization of hospital carriers and patients with penicillin-resistant strains of *S. aureus*.' E-mail to Mrs Lois Reynolds, 7 December 2007. See Gould (1958); Anon. (1958b).

Shooter–Lidwell group.[67] I was a medical student at Bart's at that time, and we dutifully had our noses swabbed from time to time, with little idea about what it was all about. They were, of course, pursuing 'hospital' staphylococci such as the 'phage type 80/81. In those studies, conducted on certain wards, they were collecting prospectively all staphylococci from all patients, from all staff and from air samples. This yielded an enormous number of isolates, and they sent them from time to time to the Staphylococcal Reference Laboratory in Colindale to be 'phage-typed. They could then put together the pieces of the puzzle.[68] In these papers, which are rarely cited these days, you can find the evidence that indicates the need for single wards; the pivotal roles were the importance of nasal carriage; the contribution of airborne transmission and of nurses to the spread of staphylococci.[69] They were able to quantify the percentage of infections that came from other patients, from the staff, and so on. The answers to so many of the questions that are asked today about the 'modern' MRSA are to be found in those papers. I reviewed these when we were in trouble with the MRSA outbreak at the London Hospital in the 1980s, to demonstrate the need for surveillance, treatment of nasal carriers and patient isolation.[70] Today, the implications of these studies for MRSA, such as the need for adequate isolation facilities and/or single wards, seem lost in history. I think if nothing else comes out of this meeting, we should highlight these unique studies, which couldn't possibly get resourced today, that describe the epidemiology of hospital staphylococci of the 1960s, which forms the basis for the successful control of the current MRSA problems, as happens, for example, in the Netherlands with their policy of actively seeking out MRSA, isolation of positive patients and energetic eradication of carriage.

Dr Georgia Duckworth: Like Ian Gould, I can't bear witness to what was going on in our hospitals in the 1960s, but I think the focus so far has been very much on England, and I think it's important to say that the MRSAs in the 1960s

---

[67] For example, Williams and Shooter (eds) (1963); Williams *et al.* (1966): 77–115; Williams (1963); Lidwell *et al.* (1970).

[68] For example, Lidwell *et al.* (1966, 1971).

[69] Professor Mark Casewell wrote: 'For example, in a multi-centre survey by the PHLS of the nasal acquisition of hospital staphylococci for 4100 medical patients in the late 1950s, there was an acquisition rate of 9.4 per 100 patient weeks. But when adult patients were nursed in modern single rooms in the newly opened Coppett's Wood Hospital [now part of the Royal Free Hospital, University College Hospitals, London], the nasal acquisition rate reduced to only 1.0 per patient weeks.' Note on draft transcript, 12 December 2007. Casewell and Hill (1986).

[70] Casewell (1986).

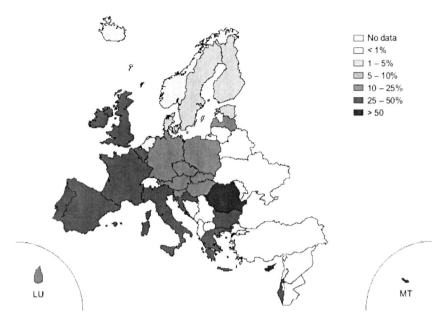

Figure 7: *Staphylococcus aureus*: proportion of invasive isolates resistant to oxacillin (MRSA) in 2005 by the European Antimicrobial Resistance Surveillance System (EARSS).

were widespread in other countries too. I remember papers from Switzerland and other European countries, including one from Denmark around 1969, which stated that around 40 per cent of all their *Staph. aureus* strains were MRSA.[71] The Scandinavian countries now have very low rates (as shown on in those international maps produced by the European Antimicrobial Resistance Surveillance System) [<1–5% in Figure 7], and we would love to be like them.[72]

So what happened? I think David [Shanson] has already alluded to the lull in the 1970s. Why did that happen? I think that's very poorly understood. Certainly the strains in the 1960s were very different from the ones that started arriving in the 1980s, like the one Mark [Casewell] mentioned at the London Hospital, which was very much like an Australian strain. But again, the first of the so-called epidemic MRSAs in the 1980s (Epidemic MRSA-1 or EMRSA-1), just

---

[71] Rosendal (1971); Shanson (1981).

[72] The European Antimicrobial Resistance Surveillance System (EARSS), funded by the European Commission, is an international network of national surveillance systems which collects antimicrobial susceptibility data for public health professionals. For further details, see www.rivm.nl/earss/result/ Monitoring_reports/Annual_reports.jsp (visited 5 December 2007).

imperceptibly went and was replaced by other epidemic strains, EMRSA-15 and -16.[73] Why do these changes occur, the lulls, and different strains coming to the fore?

Selkon: Can I bring us up to date on the evolutionary history of MRSA? Mark Enright and his colleagues have recently published a study funded by the Wellcome Trust which helps to bring the past and present understanding of the pathogenicity of *Staphylococcus aureus* together.[74] During the 1950s the most pathogenic strains of *Staph. aureus*, especially for young people, were bacteriophage types 80/81 and this was claimed to be due to their ability to produce the Panton-Valentine leucocidin (PVL). D Ashley Robinson and his colleagues have shown that the gene cassette [a collection of genes] responsible for the production of the PVL were now present in most of the epidemic MRSA clones and especially those re-emerging as community-acquired strains. Thus our old enemy returns in a new guise to haunt us.[75]

Phillips: I want to comment on this paradoxical behaviour of MRSA and, indeed, other staphylococci. First, we have had Gordon Stewart telling us about a major outbreak, while a number of us have commented that we missed the thing as it went by. We were conscious of no cross-infection, and, indeed, of no particular clinical problem. Staphylococci are like that, but, like Georgia, I do scratch my head and say: 'What on earth accounts for this?' What accounts for the fact that the clean-living Scandinavians eventually reached 46 per cent MRSA in their staphylococcal isolates, and then nothing the second-time round, when the real epidemic strains come along?[76] I think we, too, ought to be careful when we are talking about virulence and a property that I have tried to call 'epidemigenicity'. It isn't virulence, it's the ability to spread, and I think it's important to distinguish [the two]. I think those early strains were no more virulent than any other staphylococcus. And they weren't particularly able to spread, as far as I am concerned. Gordon Stewart may disagree with me, but

---

[73] Cox *et al.* (1995); Cooke and Marples (1987); Marples and Cooke (1988). See Appendix 1 with details of EMRSA-1 through -17.

[74] Enright *et al.* (2002).

[75] Dr Angela Kearns wrote: 'The paper by Robinson (which I co-authored) has been slightly mis-quoted, so the text should read: "have shown that the two genes which encode PVL were present in the epidemic 80/81 Meticillin-sensitive *Staph. aureus* (MSSA) clone and this strain has re-emerged as community-acquired MRSA following the acquisition of the *mecA* gene".' Note on draft transcript, 3 January 2008. Robinson *et al.* (2005).

[76] Rosendal (1971).

ours certainly didn't spread; they were sporadic, they cropped up here and there, and apparently were the same – all 'phage group III – and were all resistant only to penicillin, streptomycin and tetracycline.[77] Whenever anyone typed them, they all looked the same, but they behaved differently, as staphylococci often do, in different contexts.[78]

Professor Curtis Gemmell: Can I perhaps go back to the 1960s? I was a young PhD student and joined a very small group in Glasgow who were interested in looking at the virulence of staphylococci, and at that time the interest there was the search for the toxins of *Staph. aureus* that might be important in developing a vaccine. In fact we, in-house, developed purified staphylococcal α-haemolysin, coagulase and also a homologous strain of *Staph. aureus* and we thought that was the answer to producing a vaccine against *Staph. aureus*. It didn't really go very far; it did protect rabbits, but only against the whole of this strain. But in the 1960s, the change was the introduction of methicillin resistance, but people were probably side-tracked looking at why these 80/81s, and the clones thereof, were more virulent, and causing more serious infections.

If I can just say one more thing, so that when I do come back into the conversation, my real interest in MRSA started only ten years ago in 1996 when we were asked to set up the first Scottish MRSA Reference Laboratory in Glasgow, but we will come back to that later.

Stewart: In answer to that point, the staphylococcus begins in this way. You get currents of increase, and this was very marked in those early days, because in the 1950s, to which various people have made reference, there was a devastating outbreak of penicillin-resistant staphylococci due to type 52/52A and various others and then they disappeared. When methicillin resistance appeared in the 1960s, all the original pathogenic strains belonged to group III 'phage type 75/77 and so on, and then they began to disappear again.

In my review at the beginning of the meeting, I missed out some important people whose names come up incidentally in what's been said.[79] For example, Knud Riewerts Eriksen in Copenhagen was one of the first to observe that these

---

[77] Cookson and Phillips (1990); Phillips (2007).

[78] Professor Gordon Stewart wrote: 'There are cellular and morphological differences in highly-resistant strains of MRSA and also a property of regressing to a more primitive, syncytial [without cell walls] growth. See Kagan *et al.* (1964).' Note on draft transcript, 8 December 2007.

[79] See pages 4–7.

strains became quite prevalent as soon as methicillin began to be used in the university hospital in Copenhagen.[80] And, his strains were the same as ours, as were those isolated by Yves Chabbert at the Pasteur Institute, Paris.[81] They were never sent to the US, which was quite strange. For something like 20 years, right through the 1970s, there was a continuous watch in the US, and nothing turned up, and there was no problem until we had the group III strains turning up with similar virulence in the 1990s.[82] There was a carriage problem, but not a problem of severe infections. Some hospitals, some countries, missed out altogether, others were very badly affected.

Bud: I thought perhaps we should get a perspective from another country, I didn't know whether Dale Smith could say anything about how the US engaged in this from the mid-1960s.

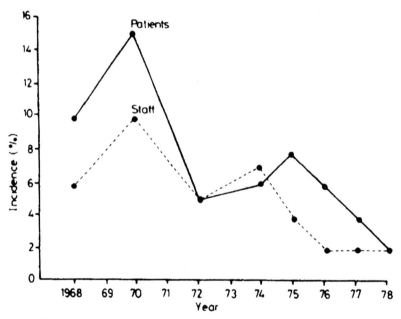

Figure 8: Nasal carriage of tetracycline-resistant *Staphylococcus aureus* (including MRSA) in a large general hospital, 1968–78.

---

[80] Professor Gordon Stewart wrote: 'Mainly of 'phage group III and this still applies.' Note on draft transcript, 10 January 2007. See Faber *et al.* (1960); Eriksen (1964); Rosendal *et al.* (1977).

[81] Chabbert *et al.* (1964).

[82] For a list of the EMRSA strains, see Appendix 1.

Professor Dale Smith: Well, you are now calling on a historian, not an eyewitness. The emphasis on the US was bifurcated. The epidemiology was being worked out by people looking at the massive epidemics of staphylococcus of the mid-1950s in neonatal nurseries and after surgery, and they were concerned with infection control much more than they were concerned with resistance or typing. It was looking at procedures and what was breaking down because of this. In the laboratories of some of the leading medical schools people were working on resistance and toxins. I know Dr Spink at Minnesota was working with toxins at that point.[83] The focus though, I think, was not on anything that was different, so much as on what had happened to surgical and obstetric training that resulted in the spreading hospital epidemics that were documented by the CDC [Centers for Disease Control in Atlanta] in the classical staphylococcus report of 1958.[84]

Stewart: Yes, Professor Smith, where you come from there is quite a problem in the spread into the community now, because some of the accounts which are most alarming now, come not from hospital spread but from the community, often via drug addicts in San Francisco.

Smith: We have a significant problem in several states. There's a bad problem in Texas, in California and it's spreading in other places into the community as well.

Professor Alan Glynn: On the question of spread and infectivity, I don't know much about MRSA. But when I was a student at UCH, over 50 years ago, Ashley Miles, the mentor of Robert Williams, used to show us the photograph of a bench of bishops and though they weren't actually picking their noses, at least half of them had their fingers very near.

Professor Brian Spratt: I would like to talk a little about the epidemiology. If you look at the *Staph. aureus* population, it's very diverse, but there are a number of very prevalent strains, and when methicillin resistance first arose, we know that the *mecA* gene moved into one of those strains, which was prevalent at the time and this was Mark Enright's work when he was in my lab.[85] If you

---

[83] Professor Gordon Stewart wrote: 'Wesley Spink foresaw the importance of enterotoxins and leucocidins [see page 30]. These infections, like those in the CDC report, were the ones that cleared with the advent of methicillin.' Note on draft transcript, 8 December 2007. See, for example, Spink (1951, 1962); Wise *et al.* (1956).

[84] US Department of Health Education and Welfare (1958). See also Ministry of Health (1959); CDC (1977): 13.

[85] Enright *et al.* (2002).

look at the very early strains during the 1960s, they are essentially all the same. What happened then was that the *mecA* gene moved into these other prevalent strains to produce a whole set of new successful strains with the *mecA* gene in them, and that I suspect correlates with the problems that we started to have with the epidemic MRSA. Now why is it that the *mecA* gene, moving into other successful *Staph. aureus* strains, gave rise to the EMRSA problem? I don't know, but presumably it is something to do with those strains being particularly well adapted to transmission within hospitals.

The other thing is that we have talked about the hospital problem and the emerging problem in the community, and it was also said that 'phage type 80/81 has been mentioned, and 'phage type 80/81 as we now know from various studies has got this PVL toxin. What Mark Enright showed quite recently in that paper in the *Lancet* that was alluded to, was that 'phage type 80/81 had now picked up the *mecA* gene and has become MRSA, so we now have a community MRSA, which has many of the properties of the old 'phage type 80/81 in causing necrotizing pneumonia and other serious conditions in the community.[86]

Shanson: Two points: first the community strains we see in the UK are not like the community strains in the US at the present time, as far as I am aware. Our experience is that they are not generally multiply resistant like the US strains and they are different types. I am not sure that anybody has shown in a British strain an 80/81 type genetic basis, and they are not usually PVL-toxin producers.

Coming back to Scandinavia and the reason that there seems to be less MRSA there than elsewhere in Europe during the last three decades, I know Rosendal and others in Scandinavia have been very keen on surveillance and strict infection control measures. The Dutch are also aggressive on 'search and destroy' policies, and I think the whole culture nationally is different from what we have seen for various reasons in the UK.[87] I am also aware that even in recent years: the British EMRSA-16 epidemic strain was transported by a patient from England to Greenland and caused a major outbreak in a hospital there, in spite of the fact that it was in danger of becoming endemic there, the aggressive actions by infection control authorities there wiped out that EMRSA-16 from that hospital, though it involved draconian measures.

---

[86] Robinson *et al.* (2005).

[87] Spicer (1984); Cox *et al.* (1995); Gould (2005). See also Department of Health (2003).

Newsom: One thing that I wonder about is how much we encouraged the methicillin-resistant strains in the beginning, because if you look in the 1960 issue of the *Lancet* announcing methicillin [Figure 4, page 11], there is also a paper from St George's Hospital, describing its use as an aerosol in the nursery. While I wasn't aware that there were problems later on in St George's Hospital, London, I did have a friend who was a medical student in Melbourne at the time when methicillin was being sprayed around the surgical wards, and sure enough ten years later Melbourne was an absolute hotbed of MRSA. So I just wonder to what extent aerosolized antibiotic contributed.

Simmons: May I suggest that the staphylococci are behaving in exactly the same way that staphylococci have always behaved, in that some are more spreadable than others, for a reason we don't know, and some are more capable of causing disease than others, often for reasons we don't know, although with some of them we think we may do.[88] If you look at the way staphylococci became resistant to penicillin, it took a period of time, then it was resistant. Tetracycline: it took a period of time, then it got resistant. Methicillin: there were some resistant ones, but it didn't get resistant so quickly. That, I think, was a critical difference in the behaviour of the staphylococcus. Some of the staphylococci had the characteristics of spreadability, which they added to pathogenicity or virulence. Some didn't.

There's not just one MRSA: there are lots of MRSAs. The staphylococcus is just picking up these things, including the characteristic of resistance to this one antibiotic, or the methicillin family of antibiotics and adding it to its library of pathogenicity factors which contribute to its ability to cause illness in modern society. The staphylococcus finds the same environmental niches in which it can rest and spring out on the unsuspecting patient who has had some nasty infection, or whose immune system is suppressed, and kills him more easily. But the organism is behaving as it always has behaved. If you look at the 1940s, 1950s, 1960s, it suddenly got resistant to the antibiotics in use. The thing about MRSA is the comparatively long delay until the emergence of widespread resistance. It didn't change the behaviour of the staphylococcus, it changed the behaviour of man, the people who were trying to treat it. If resistance had occurred within five years, we would have carried on isolating patients and doing all the things that we had done before that. But the methicillin actually suppressed our memory banks. There were people before that who were giants at that time – Reggie Shooter, Robert Williams – people that we respected and

---

[88] Parker *et al.* (1974).

who knew what could be done. How many people now could get their local authority to build a bathroom, and make surgeons shower, and then do air counts on them? They showed showers increased our counts. And yet they are still trying to get patients to shower before they have their operations.

Bud: As a historian I noticed very much at the time how people like Williams felt very beleaguered and were seeing bad practice all around them. So they were advocating the best practice, but there seems to have been an awful lot of bad practice, even then.

Dr Angela Kearns: If I can just come back to the 80/81 story, because I find it quite fascinating. I am now working in the Staph Reference Unit [Colindale] and we are looking at PVL in many different strains of staphylococci. It's found in a very diverse range of strains, as I mentioned before, but, of course, the 80/81 does have some significance for us, and, yes, we are seeing PVL-positive community MRSA, which have descended directly from the 80/81 MSSA strain. Also the US strain which was referred to earlier, is classically the so-called USA300 or what we call the ST8 strain, which we do see in this country as well. We are seeing some evidence of international spread of these PVL-positive strains with some great concern, but what we don't see is large outbreaks as yet, they are very much sporadic. It may be a 'watch and wait' situation, but that's the evidence as we see it at the moment.[89]

Professor Brian Duerden: I can only go back to the very late 1960s as a student and starting in all of this. But we have said a lot about the changing pattern, or not-so-changing pattern, of the staphylococcus, and we have hinted from time to time about the people that are involved. To pick up on what Norman Simmons said a short while ago, the emergence of MRSA had a big impact upon the medical staff, and what we saw in the late 1980s and through into the 1990s, as MRSA started to come back as a real clinical problem, just coincided with when the subject of infection was basically dropping out of the medical curriculum in many medical schools. Many of us here, as members of the Association of Professors of Medical Microbiology at the time, actually tried to survey what was happening in the reduction of teaching about microbiology and the use of

---

[89] Professor Graham Ayliffe wrote: 'In 1976/7, Edward Lowbury and Harold Lilly reported an outbreak of severe staphylococcal sepsis in the Birmingham burns unit, affecting patients and staff [Lilly *et al.* (1979)]. This was caused by a methicillin-sensitive strain, 'phage type 95, with characteristics resembling type 80/81. Towards the end of the epidemic, the relatively avirulent endemic strain of MRSA in the unit developed a similar enhanced ability to cause severe infections. It seemed likely that a virulence factor was transferred to the MRSA possibly by transduction.' Note on draft transcript, 9 January 2007.

antibiotics and infection control in our medical schools.[90] This was compounded by the General Medical Council coming out with guidance on medical education that omitted the word infection. We are still waiting to try to get that back in their recommendations. At that time when there was a neglect, then the organism was rearing its head again as a very significant pathogen.

The other people who were involved were the patients, and it is a very changing and very different patient population. I was interested to pick up, again I think it was Joe Selkon who said, in looking at patients in the 1960s that it was medical patients who came in with it and yet it was the surgical patients who were then the problem. Now many of the patients who are getting MRSA infections are medical patients, but they are desperately ill. It was the same sort of patients who were coming in with it then, but the things that are being done to them in our hospitals are effectively similar to what was being done with them on the surgical side then. So it's medical patients that are getting it, but the message is exactly the same as it was, as you described it.[91]

Dr Bilwanath Chattopadhyay: During the last four years of the 1960s I was a junior doctor in microbiology. Then I was fortunate enough to be appointed senior registrar to Professor Reggie Shooter, and to Professor Francis O'Grady, at St Bartholomew's Hospital, London. Now, somehow, we did not come across any problem in those days with MRSA, and no one can tell us that we missed MRSAs, because Francis O'Grady was there, David Greenwood as well. What happened was that I found that they were actually ruthless in recommending isolation and infection control procedures in any patient with serious staphylococcal and streptococcal sepsis. There was absolutely no excuse for not following infection control guidelines. Similarly, they also put tremendous emphasis on air-borne spread, which I think these days we are not taking much notice of. Then I joined Whipps' Cross as a consultant in the beginning of 1974. As David Shanson said, I had my honeymoon period and there was not much problem at that time, but then within two years I wanted to appoint the first infection-control nurse. My clinical colleagues thought I was a mad microbiologist, someone who wanted to take money away from the nursing budget. But again I was very lucky, because at that time one of the ward's senior nurses was no less than Christine Hancock (who later on became the Chief Executive of our Trust, followed by the General Secretary of

---

[90] General Medical Council, Education Committee (1993).

[91] See page 16.

the Royal College of Nursing). Christine Hancock, as Director of Nursing, with a newly appointed surgeon, Mr Nigel Offen, after a lot of difficulties approved the appointment of our infection-control nurse, and, dare I say, some of the senior clinicians were horrified to notice that that infection-control nurse was inspecting their aseptic practice.

Gould: I think it was Ian Phillips who asked why the Danes hadn't had a second phase of MRSA. Maybe I could be a little controversial and say it's because they learnt their lesson the first time round. Whereas I am afraid, many of our colleagues (as has already been hinted at by David Shanson) over the last 20 or 30 years, have been remarkably complacent about MRSA and, I think, continue to be complacent, even to this day. I find it very difficult to find kindred spirits when one wants to take an aggressive stance on MRSA. Our colleagues in Scandinavia find this absolutely incredible, what is going on in the National Health Service, today with the huge amount of public concern and the huge amount of resource, allegedly, put into infection control. There seems to be a complete inability to learn from the papers that Mark Casewell has already hinted at about the epidemiology of *Staphylococcus aureus* and how it spreads, and the importance of identifying cases and carriers, and isolating them properly, while taking cognisance of the problems of environmental contamination etc., etc. I could go on, but I will leave it at that at the moment.

Professor David Greenwood: When Bill Chattopadhyay remembers me at Bart's in the 1960s, I was a technician in the bacteriology lab, and, as Bill will remember, one of the reasons why we didn't find methicillin-resistant *Staph. aureus* is that we used a very antiquated method of sensitivity testing that was inherited from Paul Garrod: the test organism was spread on a blood agar plate with a glass spreader, holes were made in the agar with a cork borer and liquid antibiotics were pipetted into the holes.[92] As has been alluded to earlier, if you want to detect MRSA reliably, you not only need a better method than this, you have to use special conditions – reducing the incubation temperature to 30°C or adding a high, non-physiological concentration of salt to the agar. Moreover, there was a view prevalent at the time that a good way to keep MRSA at bay was not to test *Staph. aureus* under these strange conditions; after all, it was argued, patients weren't incubated at 30°C or given increased salt.

Shanson: I think there is perhaps one difference which the Blowers–Williams era didn't describe in relation to epidemic MRSA strains, and that is the astonishing

---

[92] See his Biographical note on pages 114–15.

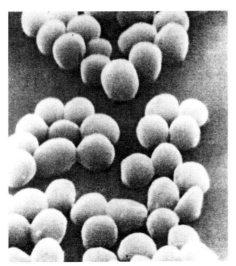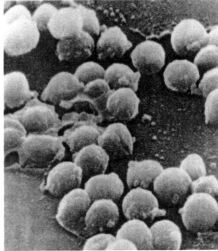

Figure 9: Two scanning electron micrographs, c. 1969, showing damage to the normally smooth cell walls of *Staphylococcus aureus* caused by penicillin. L: normal cells; R: cells exposed to penicillin (ampicillin). Source: Greenwood and O'Grady (1969): 1077.

latency that's often seen, whereby you can screen contacts of an index case and they are negative, but six months later they are positive in various sites, and that causes difficulties in the management of individuals and in the control of outbreaks. In the mid-1980s, we described the use of an isolation ward at St Stephen's Hospital, London, and although it was one of a number of measures, so you couldn't say the isolation ward *per se* helped to control the outbreak, there was no doubt that it was useful in dealing with these negative contacts as well as infected patients.[93] So that when a ward was closed, cleared of surgical patients in that ward having the MRSA epidemic strain, and the other patients in that ward were screened and were negative, we decided to also put the negative contacts in the isolation ward. This was a purpose-built ward, and indeed two of the screened-negative patients subsequently became positive. It does show the difficulties in dealing with epidemic MRSA strains, more perhaps than with the previous 1960s and 1970s antibiotic-resistant *Staph. aureus* strains, apart from perhaps the 80/81 strain which was very spreadable.

Duckworth: A query and also a point. Firstly the query: in the *Staph. aureus* story in the 1950s, much of it was concentrated on paediatric wards, particularly babies – and Graham mentioned that some of the early MRSAs occurred in paediatrics – and we also heard about the aerosolization of methicillin in the

---

[93] Shanson *et al.* (1985).

nursery at St George's Hospital. But MRSA infections more recently haven't been heavily focused on children. Bacteraemia rates have certainly been going up in children's wards, but nothing like the situation elsewhere in the hospital. Now, is that related to current strains? Or is it that practice has changed, for instance, that children now are often effectively isolated in their incubators? Moving on to my other point, this relates to what Brian Duerden was saying about neglect in medical education. It wasn't just the neglect in medical education, it was the neglect that was happening with senior management in our hospitals around the late 1980s. Many of us were using the 'search and destroy' approach then and I get very, very upset when we keep being told that we ought to do what the Dutch are now doing – 'search and destroy'. A lot of us in England were using this approach.

There were two things happening: firstly, the strains changed. We got these new EMRSA-15s and -16s, a major new onslaught on many of our hospitals, as these strains appeared to be more transmissible. It might have been to do with the epidemicity that Ian Gould alluded to.[94] But secondly, at the same time as this was happening, our ability to influence our chief executives also seemed to be spiralling downwards. Because of the financial pressures on hospitals, we were no longer able to shut wards etc. The influence of the Infection Control Doctor seemed to be waning.

Bud: What years are you talking about there?

Duckworth: I think this was towards the late 1980s. My colleagues will be able to confirm that. We established an isolation ward at the London Hospital in the early 1980s, and that has been maintained ever since, although the case for maintaining it has to be made periodically to hospital management. It was possible to establish an isolation ward then, but very difficult to set one up in the current climate. It was the sort of thing that you could do in that period. You certainly can't do that sort of thing now.

Hamilton-Miller: I would like to make two very brief points: John Pearman in Perth, Australia, managed to keep his hospital completely clear of MRSA, despite the fact that other parts of Australia were absolutely teeming with it. What he did was to admit every patient to an isolation ward before they got into hospital. The second thing I would like to say is that when the EMRSA-1 came out in 1980, we at the Royal Free Hospital, London, had the dubious distinction

---

[94] Professor Ian Phillips wrote: 'As did I, but I called it epidemigenicity (v. supra), emphasising it as a property of the organism, independent of its environment.' Note on draft transcript, 6 January 2007. See page 24.

of being one of the first places that it appeared. There was a patient who was going home on intermittent dialysis, which I thought was rather strange. We were able to get rid of the EMRSA-1 and subsequent strains, until we got to -15 and -16, where we just drew a complete blank. This reinforces the point that it's the nature of the organism, because our infection control systems have been very robust throughout the last 25 years.

Richmond: Can I change focus slightly. Arising out of all this, I do wonder whether something about the policies in relation to the funding of research in this area has not had quite a lot to do with the phenomenon that we now have. So much more work has been funded on the basic genetics, the mode of action, why antibiotics work, how organisms are resistant, and very little, or relatively little, it seems to me, in the area of how epidemics form, how the organism survives in this situation or that situation. I think it is absolutely incontrovertible that antibiotic use does influence the incidence of resistant organisms, but the coupling is very, very complex. I don't want to get off the subject, but I was always very interested in E S [Andy] Anderson's papers on salmonella, where you got enormous outbreaks of resistant salmonella type 29 and various others, and they then disappeared, even though antibiotic use didn't apparently alter substantially.[95] It seemed to me that this must point in the direction that it was something to do with the organism in relation to the host and the environment. In my experience, and maybe we will come on to this a bit later, the Medical Research Council in the 1960s and 1970s was broadly rather uninterested in that. They were all absolutely obsessed with the molecular basis of this phenomenon, the molecular basis of colonization and things like that were really very, very difficult to tackle in the way that the purists of *E. coli* K12 liked. I think there has been a very substantial shortage of funding in the area of epidemiology of infections. Now, maybe people didn't want to do the research, or couldn't do it. I know the research councils were responsive to a degree, but I think that there was also something of a positive policy to concentrate on the molecular side.[96]

Bud: Can we take Professor Richmond's intervention as a chance to make sure that in the last half hour before tea we do talk about the patterns of research, and also, so to speak, about changing the science of MRSA. I think that the line that you have taken us on following other people is something that we should spend a little time talking about.

---

[95] Anderson (1968). See also Anderson (1966).

[96]

Stewart: Mark, I wonder if you remember responding to a question raised on the nature of resistance to antimicrobials at a symposium in the US in 1967? You said: 'The realization that most Gram-negative and some Gram-positive bacteria carry extrachromosomal elements, such as episomes or plasmids, must have a profound effect on our concept of speciation'.[97] Now this was a most important point, which foresaw some of the complex mechanisms identified by Bill Hayes and others later, but did not anticipate the present problems with virulent strains of MRSA that depend on the *mecA* gene and related chromosomal elements identified in recent years in the US, Japan and elsewhere. The same genes were found as those in subcultures of the group III strains filed in Colindale during the original outbreaks in the London area, and eventually all over the world. Haemolytic streptococci, pneumococci and staphylococci possess a muramic acid–peptide kind of cell wall.[98] It doesn't apply to a lot of other organisms, which are also highly mutable.

This was 1967, but if we go back a bit to, say, 1942, when I was a house physician in the Glasgow Royal Infirmary, the problem then was not, as in Lister's day and now, staphylococcus. The problem then was the haemolytic streptococcus, because Glasgow was an incendiary place, full of shipyards, furnaces, steel works, and the big problem was burns, with streptococcal cross-infection. There was a team working in Glasgow Royal Infirmary's burns unit using crude penicillin, prepared in the Dunn School of Pathology, Oxford.[99] Their target organism was not staphylococcus, it was haemolytic streptococcus. That was my first encounter with penicillin, when I was shown those plates where this crude penicillin was inhibiting haemolytic streptococci. And then, with war wounds in view, staphylococci entered the picture, used in the armed services. And then the same sequence began to be recycled in staphylococci. One organism [strain] would come in a wave and then disappear. And in answer to the question: 'Why has there never been an epidemic of MRSA in Britain?' the answer is there is no

---

[97] Symposium on epidemiology of drug-resistant infections. Stewart G T (Convenor) Antimicrobial Agents *v.* Chemotherapy, 1967. American Society for Microbiology, 245. Richmond quote on page 293.

[98] For the characteristics of muramic acid-peptides, see Stewart (1965): 89, 92–3.

[99] Professor Gordon Stewart wrote: 'Crude yellow penicillin made in the Sir William Dunn School of Pathology, Oxford, was used in the Oxford Burns Unit for application to burned and grafted areas of skin by a team of Arthur Clark (surgeon), Leonard Colebrook (bacteriologist), Thomas Gibson (plastic surgeon), Peter Medawar (biologist and Nobel prize winner) and Maurice Thomson (house surgeon). During the same period, penicillin was cultured in gin bottles by Dr C A Green in the Royal Naval Medical School, Clevedon, Somerset, and tested in cooperating naval units.' Note on draft transcript, 10 January 2007.

room for an epidemic, because it is endemic. There's room for an epidemic in France, in Denmark, and elsewhere, and especially just now in the US, but not here, because we have had it all the time and it has become endemic, it's become accepted, and, as somebody has said, we cannot administratively deal with that under the present organization. That's quite true.

Bud: Professor Richmond, would you like to respond?

Richmond: No, all that is right. I think the point to reinforce is that these organisms are pretty clever. I sometimes wonder why it is that this organism continues to be recognizable as *Staphylococcus aureus* in view of the amount of genetic exchange that goes on.

Selkon: The most dangerous organism we have is the Department of Health. This has become very clear to me. From about the early 1990s, we were being forced into accepting excessive surgical admissions to reduce waiting lists, or to have our finances cut. I once tried to close a ward at the John Radcliffe. The Chief Executive said: 'Certainly we will do that, but you realize that it will cost us £500 000 if we don't meet our expected patient put-through rate: think of the hospital, please'. And I foolishly conceded. But the fact is that there was great pressure put on the medical profession and I am sure many others here will agree that they did not have the support which normally one would expect from the top echelons of the nursing and medical administrative staff, to be able to insist that our protection policies were adhered to. This was very difficult to accept.

Newsom: Just to go back to Georgia [Duckworth] and the type 80/81 and the babies. We had a big outbreak in the Westminster [Hospital] in 1959, I guess, and this started from a midwife with an infected finger. The babies, of course, were all kept in the nursery, so there was a seed bed for them, unlike today. The second thing is that the babies gave the germs to their mothers and 16 breast abscesses appeared. At that time if you looked for serious staphylococcal infection in the community, if you asked had they had a new baby, they very often had. So I think the babies were a definite seed bed for it all. I think the germ could be nasty to the grown-ups as well.

Spratt: The Chairman said we might talk a little bit about some of the scientific aspects and Mark Richmond has already mentioned that a lot of money was going into mechanisms of resistance in the 1960s and 1970s. I thought I might just look at the history of what we know about the basic mechanism of resistance to methicillin in *Staph. aureus.*

Soon after the discovery of MRSA, as Mark has alluded to, there were some curious properties about the instability of methicillin resistance, and the possibility was raised that perhaps it was on a plasmid and maybe β-lactamase was somehow involved. I think it was probably Keith Dyke in Oxford in the mid-1960s, who, I think, showed convincingly that you can get rid of the β-lactamase plasmid, but you don't get rid of methicillin resistance.[100] I think that pretty clearly showed that the mechanism had nothing to do with the β-lactamase plasmid. And really there was not a lot that had progressed in terms of the mechanism of methicillin resistance throughout the 1960s and 1970s. There were some suggestions that perhaps it was something to do with the cell wall. It became gradually clear that resistance was due to a chromosomal gene, but there was a very poor understanding of this, throughout the whole of the 1960s, about the precise mechanism of action of methicillin, or of penicillin and the β-lactam antibiotics in general. It wasn't until 1975 when it was possible to identify in cell wall synthesis, in peptidoglycan synthesis, the individual enzymes that were inhibited by penicillin.[101] It then became possible to look at the interaction of penicillin with each of the individual enzymes in cell wall synthesis that were inhibited by penicillin, which became known as penicillin-binding proteins (PBP), and it became clear towards the late 1970s that although resistance to penicillin in most bacteria was due to the β-lactamases, there were some examples where resistance was due to alteration in the target enzymes, alteration to PBPs, and particularly, for example, in the pneumococcus, where all resistance to penicillin is due to alterations in the PBPs. Alex Tomasz was particularly involved in this.[102] But in terms of MRSA, the breakthrough with understanding the mechanism of MRSA was actually in Cambridge, and it was Derek Brown and Peter Reynolds who in 1980 looked at the PBPs of methicillin-susceptible and methicillin-resistant *Staph. aureus* and showed that there was what appeared to be a new penicillin binding protein which was highly resistant to inhibition by methicillin and other β-lactam antibiotics.[103] They hedged their bets slightly, they weren't quite clear whether it was a large amount of one of the existing PBPs, or a new PBP. But then work, particularly from Barry Hartman and Alex Tomasz in the next few years showed very clearly that MRSA were resistant because they had somehow got

---

[100] Dyke (1969).

[101] Spratt (1975).

[102] Hakenbeck *et al.* (1980).

[103] Brown and Reynolds (1980).

hold of an extra PBP, which was resistant to essentially all β-lactam antibiotics.[104] That's probably still a unique mechanism of resistance to penicillin. Slightly mysterious, where did this resistant PBP gene, which is called PBP2´ [known as 2 prime], come from? Of course, it's encoded by the *mecA* gene, and it was Song in Matsuhashi's lab who managed to clone the *mecA* gene and showed that MRSA strains, or the first MRSA strains had quite a large piece of DNA which included the *mecA* gene and a number of other genes, including some encoding other drug resistances. Song and Matsuhashi suggested that the *mecA* gene had somehow come in and this penicillin binding protein gene had fused with the β-lactamase gene to produce a PBP which is under the control of a β-lactamase.[105] That is more or less where we are with MRSA. An unusual mechanism, a new PBP coming in, resistant to β-lactams. We still don't really know where it came from. There are some suggestions where it came from, but I don't find them terribly convincing.

Simmons: Going back to Professor Richmond, who asked how people behaved towards funding research. One of my colleagues, the late David Williams, had a research project in 1961, all written up, ready for application for a research grant into the transmission of staphylococcal infection, and it was all favourably received, but when methicillin came out within a year, it was rejected, because it wasn't a problem any more. Secondly, the incidence: everyone is saying that children were very important. In 1960 the incidence of staphylococcal post-operative sepsis in gallbladder operations was 14.9 per cent, in breast operations it was 12.7 per cent, in varicose veins, hardly a paediatric disease, it was 9 per cent. So the idea that *Staph.* infection occurred only in children is in the eye of the beholder. Now, MRSA doesn't exist in a little capsule, at the moment anyway, it exists principally in our hospitals, and what has changed there is the system of patient management. The present system of patient management is designed to bring about a reduction of waiting times, which is perceived to be the most important issue. If the prevention of infection was the most important issue, we would control the infection. It's a question of priorities.

Dr Elizabeth Price: Bill Newsom mentioned that outbreaks of *Staph. aureus* were not uncommonly reported from neonatal units in the 1950s. One of the reasons that these do not seem to occur to such a large extent now is that infants are usually 'roomed-in' with their mothers. This means that the baby remains

---

[104] Hartman and Tomasz (1984).

[105] Song *et al.* (1987).

at the side of the mother's bed, rather than going to a large nursery with the possibility of coming into contact with other infants, their mothers and members of staff, all of whom could transfer infecting organisms, including staphylococci. Another reason is that most mothers and babies leave the maternity unit because they are discharged very rapidly. As a result there is less chance for an infant to acquire infecting organisms in hospital.

Joe Selkon mentioned the relevance of poor ventilation in hospitals in the 1960s. Well, that situation may be even worse now and I am talking about the late 1990s onwards. In hospitals with natural ventilation, we are not allowed to open windows more than 10cm [100mm] because of the possibility that patients might fall out or attempt suicide by jumping out. This is an NHS Estates Health Technical Memorandum requirement.[106] I think we are ending up with sweaty patients and perspiring doctors, and that is a great way of transferring organisms. I would like to see more research on organisms and the environment, particularly looking at humidity and heat. We have lots of computers and heat-producing equipment as well as poor ventilation in our hospitals. Maybe in our research we should be looking at the hospital environment and its effect on infecting organisms.

Shanson: I would like to ask a research question following some personal observations while working for the last four years in a children's hospital. Most weeks there are two or three new MRSAs detected, but they usually don't spread and the patients are promptly isolated anyway. But over the last two years there was one particular strain seen where we experienced extreme difficulty in controlling its spread. We gave the strain the name of the index patient.[107] This strain had a very characteristic multi-resistant pattern, and it particularly affected immuno-compromised children with skin lesions. Later it also infected other children without skin lesions. It was noted how persistent this organism was in single rooms, even those which had been cleaned once, twice or occasionally three times. It sometimes needed hydrogen peroxide treatment to eradicate it

---

[106] *Health Building Notes* and *Health Technical Memoranda* were published by NHS Estates until its duties were transferred to the Department of Health Delivery Group on 30 September 2005. *Health Technical Memorandum no. 55* covered the requirements for windows as part of a series on building components, which originally appeared in 1989, with a second edition in 1998.

[107] Dr David Shanson wrote: 'This outbreak problem with a difficult spreading strain in the children's hospital was not published, although further characterization of the isolates was carried out at the Staphylococcal Reference Laboratory at the HPA Laboratories at Colindale.' E-mail to Mrs Lois Reynolds, 5 December 2007.

from the environment.[108] So it seemed that this particular very virulent strain resisted drying and I wonder whether anybody is doing research to look at the most spreadable strains, if you like, in terms of their resistance to drying, and whether they also compare it with the other strains at a genetic level in relation to drying.

Gould: Norman Simmons correctly said that the focus of MRSA in the UK is still the hospital, and yet paradoxically the greatest single barrier to hospitals controlling MRSA at the moment is that there is now a large, dormant, asymptomatic population of MRSA-carriers in the community, who have picked up their MRSA in the hospital. Now, we recognize that within the UK overall, probably between 5 and 7 per cent of hospital admissions are asymptomatically colonized with MRSA. Just to bring you bang up to date, a recent health technology assessment (HTA) from Scotland has costed a national programme of admission screening for everybody and isolation and decolonization based on mathematical models which originally emanated from Ben Cooper and his colleagues in London.[109] The HTA reckoned that per annum an average-size teaching hospital with 700 or 800 beds would have to spend £17 million on this process, but within five years would have brought the underlying MRSA rate down from 7 per cent to 1.4 per cent. But, of course, the possibilities of getting such funding in the present circumstances must be open to extreme doubt.

Gemmell: I am currently the Director of the Scottish MRSA Reference Laboratory and one of the things that we are doing at the moment in our research is that we are looking for genes of virulence among MRSA strains and particularly strains from bacteraemia. We have looked at something like 200 strains over the last two to three years and one thing we are detecting is that there are strains out there, on the MRSA, in which the virulence factor is quite a potent one, such as TSST1, the toxic shock syndrome toxin. We were finding some that produced the β-haemolysin, an old-fashioned toxin which was historically always associated with animal strains, and we are also seeing occasionally PVL toxin and also epidermolytic toxins and it's called the scalded skin syndrome toxin. So the toxins are there among the MRSA, but they are very spread out among different sub-clones.

Ayliffe: Can I comment on grants? There used to be a MRC research committee which had a fair amount of money available, a vast sum was spent on a multicentre

---

[108] See, for example, French *et al.* (2004).

[109] Cooper *et al.* (2003). For the October 2007 version, see Ritchie *et al.* (2007).

trial for hip replacement surgery in the 1980s.[110] Then, suddenly at the end of that, the MRC stated that, 'We don't need it any more'. Individuals had to send in their own research grant applications for projects of two or three years, which was quite different from putting a lot of money into a large research scheme.

Bud: Do you remember when that was?

Ayliffe: I think the early trial was in about 1980, and it must have been some time after that.

Chattopadhyay: We have heard about preoperative screening. With a reasonable amount of funding, we opened the first of our MRSA wards in 1989, and we could manage to control that outbreak at that time, and there was a dedicated SHO/Registrar for that unit. Second, because of the problems in orthopaedics in the mid-1990s, we had three wards: one was the designated MRSA ward, one was the very clean ward for prosthetic implants and the third was for other categories of patients. That was very successful until the authorities decided that there were too many beds in orthopaedics, so they closed one of the wards, and the whole thing fell apart. Over the last nearly ten years, we have an isolation ward – not just an MRSA ward – with 17 beds, nine cubicles and two bays with four patients in each, where we admit MRSA-positive patients plus *Clostridium difficile* and so on. Patients with open pulmonary tuberculosis are admitted to an isolation cubicle with en-suite facility. It is not the solution to the problem, because there are several disadvantages: one is the mixing of disciplines, which the consultants don't like. They don't like their surgical patients to be mixed with the care of an elderly medical patient with a large leg ulcer, carrying MRSA. Secondly, it is not feasible to transfer the MRSA-positive patients from intensive care unit (ICU) or orthopaedic patients on traction, or from paediatric units. Thirdly, it is very expensive to run with regards to cost of drugs used and also bed occupancy and nursing staff ratios to patients. It is very difficult to recruit nursing staff, because they don't find it very rewarding to be a specialist nurse in infection control; they would rather do surgery – vascular, orthopaedics or something. And lastly, the junior doctors tend to forget the outliers [patients not admitted to their designated wards], a patient on that isolation unit. So the patient may not be visited by the team. There were advantages as well from the point of view of quality of isolation and the control of infection, which was excellent.

---

[110] Lidwell *et al.* (1982, 1984). For a Cochrane Review of antibiotic prophylaxis, see Gillespie and Walenkamp (2001).

Training was excellent, obviously with developmental expertise. But this is not the solution, because I can tell you from my experience that it is very expensive to run. You cannot have a bed occupancy in that ward of 80 or 90 per cent. If your incidence of MRSA goes down, you can't fill in those beds, and you are under tremendous pressure from the administration to admit patients into that ward, when there is not a single bed available anywhere in the hospital. Likewise, whenever there are more than 17 patients with MRSA infection/colonization, all of them could not be accommodated in the isolation ward.

Duerden: To pick up your point again about research and research funding and the pattern of research: we heard from Brian Spratt of the excellent work that has been funded on the mechanisms of resistance and the work of Mark Enright, who described the genetics, but very little money has gone in over the past 15–20 years at the applied end of this, at the interface between good research science and clinical applicability and infection control. There has been very little funding there; it's about two weeks since I was talking to some colleagues at the MRC – fortunately they are actually realizing that this had happened and the question they were asking was whether this had an impact. Well, the obvious answer was 'Yes': a huge impact, because over the same period it was coupled with the various research assessment exercises (RAE).[111] I have no problem with wanting to assess the effectiveness and the quality of university research, and research in the medical schools, but it did go quite against what had been the traditional pattern in medical schools of the clinical academic, who was both an academic and doing clinically applied work in the hospitals. Again, I look around the room and think of the number of clinical academics who are represented here. I would be hard pressed to actually balance their number with currently active clinical academic microbiologists, who should have been their successors. It's a diminishing number and these are the people who should have been carrying on the work of Shooter and Blowers and colleagues then, and the people who are here. This has not been an area that has been attractive to the major funders and we are seeing the results of that now. I hope that the recognition might turn round this issue, but we have got to find the people who are capable of doing the work now.

Duckworth: I would like to comment on Bill Chattopadhyay's point – the litany of all the disadvantages of an isolation ward. We faced them all in David Williams' department when we were setting up the isolation ward at the London

---

[111] The Research Assessment Exercise was first undertaken in 1986. For further details, see www.rss.org.uk/main.asp?page=1224 (visited 30 January 2008).

Hospital in the early 1980s. Sometimes it seemed an extraordinarily uphill struggle, with surgeons not wanting their patients transferred there etc. But I really thought we had won the day when one of the general surgeons came up to me and said that he wished all his patients got MRSA because they received so much better care on that ward. We had orthopaedic patients in traction there, we even had some ICU and paediatric patients. It was just an ordinary ward, nothing specially built. It had cohort areas and some single rooms. When we didn't have enough MRSA patients to fill it, we were able to put patients with other infectious diseases in the side rooms. Infection control practice got so good on that ward, that we did not get spread of MRSA to the patients being isolated for TB or chickenpox or other infections in the side rooms.

I would also like to comment on Bill's point about expense, because I was involved in an Health Technology Assessment (HTA), which included a systematic review of the effectiveness of isolation in the control of MRSA and some mathematical modelling. The modelling showed that even if the isolation ward was not big enough for the size of the problem in the hospital at that particular time, it still had an impact on ultimate numbers of infections.

Richmond: Just one comment on this funding business, a personal comment. When I moved to Bristol in 1968, I put a large grant application in to the MRC to work on staphylococcus and it was turned down. The message came through that we will fund you if you put in for a grant to work on *E. coli*.

Selkon: The cost of looking after 200 patients with MRSA in a Belgian hospital is put down at £2.6 million.[112] When we talk about costs, it's not just the costs of what you have to do, it's the costs you will incur if you don't do it, in terms of isolation of source cases. Very seldom do you have a chance to have a proper debate upon the total costs involved with the disease and the costs for running an isolation unit.

There is another point I want to make in terms of what has just been said. We must realize – and we will see this in the paper from Oxford published in the *British Medical Journal* this week, which shows that we have a major problem of patients coming back into hospital with MRSA bacteraemia during their hospital stay or on readmission after they have been discharged.[113] The mortality

---

[112] Dr Joe Selkon wrote: 'For cost comparison, see Abramson and Sexton (1999).' Letter to Mrs Lois Reynolds, 25 June 2007. See also Pirson *et al.* (2005).

[113] Wyllie *et al.* (2006).

rate in these patients was 29 per cent. Let's get our game together and look at hard facts, then you will see that the cost of isolation is, in fact, relatively smaller than the real cost, if we have to pay damages on human life, as you would in a motorbike accident.

Simmons: Why are the Dutch better than we are at this? Well, I think they have learnt that the time to stick your finger in the dyke is before the flood.

Gould: A point on cost, following on from what Joe Selkon said: I am afraid that in this day and age most hospitals are actually bankrupt and it's impossible to get an extra penny to spend on infection control. Even with the promise that you will repay a thousand-fold in the following year, the money just won't come.

Dr Stephanie Dancer: I have had a wonderful afternoon so far, and I am sure it's going to continue. It's been fantastic to hear so many relevant observations from the times of the staphylococcal pioneers, whom I hold in great respect. In answer to Mike Emmerson, who had suffered the indignity of a swab shoved up his nose in the 1960s. I had that too in the 1980s undertaken by Professor Shooter, and I didn't know why at that stage either. In the late 1980s I defected across the Thames to Guy's, where I did a thesis on toxin-producing staphylococci, and at that point in time MRSA wasn't big business, it was toxin-producing *Staph. aureus*. Little did I know that the work that I did then on the epidemiology of *Staph. aureus* would be so relevant to the present time, where I am a consultant at a busy teaching hospital in Glasgow and fighting MRSA every day, two positive bacteraemias a day, and, in order to juggle patients about, cohorting positive patients on the ward, because we have no isolation facilities and indeed no spare beds at all.

My final point, said with a huge amount of passion, is that I have just spent the last two years trying desperately to get some money together to look at hospital cleaning, because it's something about which I feel very strongly. I went through all of the usual sources, and jumped through all the hoops and got through to the final stages with at least three legitimate funding bodies, only to be turned down for various interesting reasons. I finally accepted the offer of money from the health union, UNISON, to set up some cleaning trials in a surgical unit of my hospital, where MRSA is endemic. That trial is going to start next week.[114]

---

[114] Dr Stephanie Dancer wrote: 'The trial finished in August 2007, and the first paper, "Monitoring environmental cleanliness on two surgical wards", has just been submitted for publication; the second will be entitled, "What is the impact of targeted cleaning on methicillin-resistant *Staphylococcus aureus* (MRSA) in a hospital?". See White *et al.* (2007, 2008); Dancer (2008).' E-mail to Mrs Lois Reynolds, 16 January 2008.

Bud: If you would just like to turn over your exam papers [to page two of the Witness Seminar Programme], there are a series of issues which I am sure we will cover this afternoon. We need to cover something more about the pharmaceutical response and we are really very fortunate that Robert Sutherland, a member of the Beecham team from the early days, is with us. We need to talk, I think, a bit more about surgery and the changing attitudes to infection. I hope that we will talk a little bit more about epidemiology and the hospital response, and then finally something about the public response in the 1990s. So, I think there is a series of issues on the second page [of the programme], which will provoke a valuable response. We should begin with Robert Sutherland, who will get us to think about the pharmaceutical response to the challenge of resistant organisms.

Dr Robert Sutherland: I feel a bit intimidated, being the sole representative of the pharmaceutical industry here today, along with leading academics. At the previous Witness Seminar on 'Post Penicillin Antibiotics' in 1998, I think due credit was given to the role of Professor Sir Ernst Chain in leading to the development of the discovery of 6-aminopenicillanic acid (6-APA) and, of course, the development of semisynthetic penicillins.[115] When the Beecham workers went to work in Professor Chain's fermentation laboratories in Rome, it was not with the intention of discovering the penicillin nucleus, but the fermentation of p-aminobenzyl penicillins that could be chemically modified. And, of course, the story of the observation of the discrepancy between the chemical assay and the biological assay is pretty well known.[116] I think it is serendipity in a sense, in the same way that Fleming's discovery of penicillin was a piece of serendipity. But, of course, what is interesting is that other companies, Merck and, I think, Eli Lilly had also made the same observation, but it was the Beecham workers Rolinson, Bachelor, Doyle, and Nayler, who explained the difference between the assays and which led to the isolation of 6-APA at Brockham Park, Surrey.[117] Professor Chain was a very enthusiastic consultant and did drive a lot of the research in the early days.

The first penicillin to reach the market was the semisynthetic penicillin, phenoxyethyl penicillin, *Broxil*, which was just a modified penicillin-V, and

---

[115] See Tansey and Reynolds (eds) (2000); Rolinson and Geddes (2007). See also Abraham (1983). A collection of Chain's papers, CMAC/PP/EBC, is held in Archives and Manuscripts, Wellcome Library, London.

[116] Ballio *et al.* (1959). See also Lazell (1975).

[117] Dr Robert Sutherland wrote: 'I believe it is important to clarify that the isolation of 6-APA was made at Brockham Park and not in Rome.' Note on draft transcript, 23 January 2007. See Figures 1 and 10.

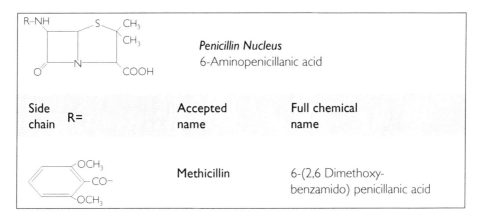

Figure 10: The structure of the penicillin nucleus (6-APA) and the methicillin side chain. Source: Tansey and Reynolds (eds) (2000): 62–3.

caused some excitement among the commercial colleagues, but very little among the scientists. But it was the discovery in 1959, not long after the isolation of 6-APA, of methicillin, with the brand name 'Celbenin', a 6-dimethoxyphenyl penicillin, and its stability to β-lactamase and consequently its activity against β-lactamase-producing staphylococci, which caused great interest, because the potential to treat these penicillin-resistant strains of staphylococci or multi-antibiotic-resistant strains of staphylococci was very exciting. In the early days, of course, the papers were talking about 'Celbenin'-resistance to staphylococci, before the approved name became methicillin. But, I now see in a recent paper in the *Journal of Hospital Infection* that we are talking about meticillin-resistant staphylococci.[118] So, in a sense, we are going round in a big circle. Whether meticillin will catch on, instead of methicillin, will be quite interesting. I think it was Graham Ayliffe who said: 'Of course, nowadays methicillin is off the market, no longer sold and younger members of the medical profession won't know about it, so they will probably be very happy with meticillin.'

However, after the discovery of methicillin, intensive work was done, obviously, on β-lactamase-stable penicillins. In the Beecham Laboratories, several hundred strains of *Staph. aureus* were tested for susceptibility to methicillin in the early days and, as I replied to Ian Phillips, I don't think any resistant strains were missed because it was traditional to test these staphylococci at two inocula, a large, undiluted inoculum, and a diluted inoculum, analogous to the testing of

---

[118] See, for example, Coia *et al*. (2006). Meticillin is the accepted spelling used in the WHO's 2006 *International Pharmacopoeia* (4th edn) guidelines.

benzylpenicillin against *Staphylococcus aureus*, where β-lactamase is the problem. So, I don't think we missed any strains.

We did notice the heterogenous resistance early on. I think that it was described in the first paper in the *British Medical Journal*; we did look at it. What I found of interest, was when we were doing further work, where we subcultured β–lactamase-producing MSSA in the presence of methicillin, we could select resistant cultures with this heterogenous resistance. What was interesting to us I think, was that when you subcultured these strains back into antibiotic-free medium, the strains retained this minority population of resistant cells, whereas when the same kind of test was done on other antibiotics – tetracycline, streptomycin – the cultures reverted to full susceptibility in the absence of the antibiotic. Very briefly, after the discovery and development of methicillin, we then went on to other β–lactamase-stable penicillins. The first was oxacillin, which we at Brockham Park concluded was not really sufficiently stable to staphylococcal β-lactamase, nor were the oral absorption characteristics sufficiently favourable, but this compound was taken up by Bristol Laboratories [Syracuse, NY] with whom we had a licensing agreement, and was very successful in the US. Next in the isoxazolyl series was cloxacillin, which we felt had the degree of activity and β-lactamase stability that was required, and was well absorbed orally and by injection. And, of course, that was eventually followed by flucloxacillin. We did examine many compounds and we could not find any that we felt showed real improvement on flucloxacillin or cloxacillin. The chemists thought that they had that area sewn up, but, of course, when Boots produced quinacillin, this was a great blow to their pride.[119] As an anecdote, we had a visit in 1967 when the testing of methicillin-resistant staphylococci, of course, was a problem, certainly the inoculum size was very important and the observation that incorporation of 5 per cent sodium chloride in the agar made it so much easier. We had a visit in 1967 from Dr Douglas Annear from the Royal Hospital in Perth [Australia]. He was showing us his results on the effect of temperature on the detection of the methicillin resistance. The higher the temperature the less resistant the organism became up to 43°C. And, of course, at 30°C, resistance is much more readily detected. He then astonished George Rolinson and me by announcing that (if my memory plays me right) he had been either to the NIMR at Mill Hill, or to Colindale, where he showed these results to some unspecified person who told him that they couldn't be correct, because somebody else would have reported them a bit earlier. So, not exactly a

---

[119] Boots' quinacillin was introduced in 1963. See Richards *et al.* (1963).

prophet in his own country. In our work at Brockham Park we never found any compounds active against methicillin-resistant staphylococci, apart from one of Professor Chain's compounds.

This story started with Professor Chain, so I will finish with him. He and his workers at Imperial College isolated the antibacterial agent which he called pseudomonic acid from a strain of *Pseudomonas fluorescens*, well known to produce antibacterial activity. I think it had been reported by Garre in the 1880s. Professor Chain isolated the material and he was very enthusiastic, until one day he seemed very depressed about the whole subject of pseudomonic acid, now known as mupirocin.[120] When we inquired why, he said: 'Well, of course, it's a shame that the compound is so toxic, with haemolytic activity against red blood cells'. What he was confusing was the high serum binding of the compound, which was a disadvantage, but the main problem was that, although mupirocin was well absorbed in the body, orally and by injection, it was extensively degraded to inactive monic acid. So, it had no potential as an oral or parenteral agent in the early 1970s. Much later in the 1980s, somebody had the idea of suggesting it as a topical agent, and I believe that topical mupirocin ointment is now used for the treatment of the nasal carriage of MRSA. So, the wheel has gone around, starting with Professor Chain and finishing with him.

Emmerson: About this time Beecham's released mupirocin (formerly called pseudomonic acid) in a polyethylene glycol base.[121] We in north London had a problem with a major outbreak of MRSA infections on a urology ward and, since mupirocin was a bit experimental, thank goodness I was a fully paid up member of the Medical Protection Society at the time. Through our surveillance, we found that the main culprit causing the outbreak in a core ward of the hospital was the single remaining urologist. I was a bit nervous about sending him off for the usual three weeks of cetrimide and convalescence in the Swiss mountains, because I thought that if my prostate played up, I would have despatched my only hope. So, we swabbed him and found that his nasal carriage sites and axilla yielded pure growths of the outbreak strain. We had this gel and, of course, we had to tell him that if you put it up your nose it will melt (as it is in a polyethylene glycol base similar to antifreeze). I couldn't send him off work so we had him on pseudomonic acid, so to speak, for 36 hours, which miraculously cleared him of the MRSA colonization, and we were

---

[120] Chain and Mellows (1977). See also Fuller *et al.* (1971).

[121] See note 128.

able to keep him on this treatment, while we coped with the rest of the other colonized/infected nine patients. A few years later I moved from north London to Belfast; we had a major problem with neonates, and with a slightly different preparation, so we started treating the ultra-low birthweight neonates with the same compound. We had to be careful to make sure that no pseudomonic acid was carried over into the recovery growth medium, otherwise this would have biased our results. We overcame this problem by using some activated charcoal and sheep red cells etc. At that outbreak, too, we were able to manage, but it was at a time when, as Robert [Sutherland] says, pseudomonic acid was absorbed into the blood stream, broken down into monic acid very quickly, and caused none of the side-effects of hexachlorophane or some of the other compounds. Also we were guessing on the dose, and when Mark Casewell came around and started doing dose response curves, we realized that we were being rather zealous in our management, but at least it broke the two early major outbreaks.[122]

Duckworth: I am surprised Mark hasn't put his hand up, so I'll comment! At about the same time that Mike Emmerson was referring to, we were also doing work on mupirocin (or pseudomonic acid) under Mark Casewell at the London Hospital. We started using it as part of the control programme. We had just opened an isolation ward and were screening around four wards a week. This was an enormous 'search and destroy' operation.[123] Mupirocin made a tremendous difference to our control efforts, as, prior to that, any staff who were found to be colonized were off work for ages. We tried clearing nasal carriage with chlorhexidine creams (even using the high concentration cream, which I think was the vaginal preparation), but were not successfully clearing carriage quickly. Then along came mupirocin and not only did it clear carriage of MRSA, but it cleared it so fast that it was no longer necessary to send simple nasal carriers off work. After all, it had already taken a couple of days to find out that they were positive, so another day for clearance of carriage would make little difference to the risk of exposure on the ward. So, it had a dramatic impact on control at that time.

Ayliffe: May I mention a controlled trial that Edward Lowbury did in the Birmingham Burns Unit?[124] Most of the burns patients there in the 1970s were colonized by MRSA and there were virtually no clinical infections at all. He

---

[122] Casewell and Hill (1989, 1991).

[123] Duckworth *et al.* (1988). For work with mupirocin, see Casewell *et al.* (1984); Hill *et al.* (1988).

[124] Lowbury *et al.* (1977). See also discussion on page 30.

treated carriers with flucloxacillin for four days. Surprisingly, the MRSA were actually removed from half the burns, and none were removed from the controls. So, the explanation was rather difficult, and it was suggested that when burns had got all their dressings on, the temperature at the burn surface was probably 37°C. Further clinical studies on this observation were not made.

Gould: Two contemporary points about mupirocin: in common with many other antibiotics, there is certainly a shortage in the UK, if not a world shortage, and one wonders, as its use expands, and it undoubtedly will, whether production will be upgraded. But as I say, this is a recurring problem with many antibiotics these days. But the other note, maybe more cautious, is that a Cochrane Review two or three years ago concluded that there was no good evidence that mupirocin should be used routinely in attempting to decolonize patients with MRSA.[125] This is, perhaps, a somewhat surprising finding, because I think everyone in this room will have anecdotal evidence of the success of mupirocin.

Casewell: Yes, I was surprised that mupirocin initially received such a poor reception.[126] I presented the results of our *in vitro* work and the first clinical trial to the Interscience Conference on Antimicrobial Agents and Chemotherapy (ICAAC) in the US in 1987, but there was very little interest from the Americans.[127] They seemed less aware that the spread of MRSA might have to do with nasal carriage. Coming to the present time, again I was surprised that the Cochrane method of reviewing research and producing guidelines for MRSA control recently came up with very dubious conclusions about the use of this compound.[128] I think that view will be revised in future and casts doubts on this Department of Health-inspired mechanism of interpreting scientific literature.

In terms of the supply of mupirocin, it was pretty hard in the 1980s to persuade the company [Beecham Pharmaceuticals] that this was a potentially exciting and marketable topical agent that had come at just the right moment for MRSA

---

[125] Loveday *et al.* (2006).

[126] Professor Mark Casewell wrote: 'It was the impression that they didn't think nasal carriage was pivotal in the control of epidemic staphylococcal spread and were therefore (disappointingly for me) excited by the prospects of using mupirocin for the elimination of nasal carriage of MRSA.' Note on draft transcript, 12 December 2007.

[127] *Program and Abstracts* of the 27th Interscience Conference on Antimicrobial Agents and Chemotherapy, abstract no. 173, by the American Society for Microbiology, 4–7 October 1987, in New York, NY.

[128] Loveday *et al.* (2006). For a WHO review of recommendations by expert policy groups, see Avorn *et al.* (2001).

control. In 1984, when we had an outbreak of MRSA at the London Hospital, Beecham's came to my lab and wanted to market it for infected skin lesions, for the dermatologists. I asked them whether they had heard of MRSA? The representative said: 'No', so I explained that we were witnessing one of the first outbreaks of MRSA in the UK, and I suggested that, used intranasally, it might treat carriers, a great advance as we were all desperate to find something to clear nasal carriers. Then they did take an interest. It is available now in the US, although they may still need to go to the early literature to appreciate the pivotal role of nasal carriage in the spread of MRSA.

Selkon: I am surprised that we actively treat burns patients for *Staph. aureus* colonization. I was trained in the opposite direction by a plastic surgeon of great experience in Newcastle who said: 'Leave the *Staph.* alone, it helps the graft to stick'. I worried about this, so I went to Ted Lowbury in Birmingham, put the case to him, and he pondered for a while and said: 'I think there is something in that, I wouldn't treat a *Staph.* on a burn unless there was actual pyrexial evidence of infection'. I have always followed that advice and I think we have to be very careful when we prescribe antibiotics for organisms, on the surface, unless we do so with good evidence that it is necessary to do so. They may not be playing a pathogenic role.[129]

Stewart: If you forgive me making a comment, that's what I referred to when I talked about the work in the burns unit in Glasgow, because there were all kinds of injuries and burns from war factories and so on. *Staph.* was not a problem, it was the superficial burns, the haemolytic strep that was a problem, but the *Staph.* was not, and I think people there – Leonard Colebrook, Peter Medawar, Tom Gibson and others – were perhaps subconsciously observing that haemolytic streptococcus was a major problem.[130] Now, if you think about that and translate it into the world of the maternity ward, the main cause of maternal mortality was sepsis due to predominantly haemolytic streptococci; it was never a staphylococcal infection, although many of those women were carriers of pathogenic staphylococci in the vagina or elsewhere. What really made a difference then was something that we don't have now: a committee was appointed, but unlike contemporary committees it wasn't part of a multiple

---

[129] Professor Gordon Stewart wrote: 'It is important, therefore, to check polymorph response, erythrocyte sedimentation rate (ESR) and C-reactive protein (CRP) in clinical assessment of superficial infection, and to check the phenotype and genotype of the *Staph. aureus* in the lesion and nose of the patient and attendants.' Note on draft transcript, 8 December 2007.

[130] See, for example, Lowbury (1983); Simpson (1988); Clark *et al.* (1943).

committee resistant to organization, it was a single executive committee and maternal mortality dropped from being quite high and a problem, very quickly, within a few years in the 1950s.[131]

Simmons: I would like to stress something that Dr Sutherland said. Methicillin was discovered in 1959 – the Jevons paper was published in 1961 – and it was in general use by 1961. I am told that there are over 90 agents effective against methicillin-resistant staphylococci that have not been pursued, because it would take at least ten years to bring them to general use. Question: are we right to apply the same regulations to the use of a new antibiotic and its production for general use, as we do to the use of other drugs which are used for much longer than an antibiotic would be?

Ayliffe: May I comment on what Joe said? I wasn't actually suggesting that cloxacillin was being used for the treatment of the staphylococcus. It was used for streptococci and it was an accidental finding that MRSA were removed and followed up with a trial.[132]

Professor Gary French: May I comment on antimicrobial resistance and new drugs for MRSA? Not only are these organisms methicillin resistant, but they are often – not always but often – multiply antibiotic resistant. One of the reasons is that the chromosomal area where the *mecA* gene is inserted has other sites where other variance genes can be inserted, making them stable. These organisms also rapidly become resistant to quinolones by chromosomal mutation. About 90 per cent of MRSA strains are now resistant to quinolones, while methicillin-sensitive *Staph. aureus* (MSSA) generally remains fully sensitive to quinolones. Thus 'multiresistant' *Staph. aureus* is a better phrase than 'methicillin-resistant' *Staph. aureus* and their spread is probably encouraged by antimicrobial therapy in hospitals.

Most of the antibiotics used against *Staph. aureus* are not as good as methicillin/ flucloxacillin. Flucloxacillin is a great drug: it penetrates the cells and tissues well, is highly bactericidal and is one of the great drugs of the twentieth century. Many of the other agents that we have for *Staph. aureus* appear not to be so effective (although older agents have not been trialled against MRSA). Once methicillin resistance emerged, we were forced back on older drugs, especially vancomycin, a drug which is only available parenterally [using a muscle, vein

---

[131] The local sanitary authority was to send a medical officer of health to be a member of the control of infection committee set up by hospital management committees and local hospital boards. Ministry of Health (1956): 8. See also Barber *et al.* (1960).

[132] Lowbury *et al.* (1977).

or any pathway in to the body other than by mouth], is potentially toxic, only slowly bactericidal and has relatively poor tissue penetration. For 30 years there was no resistance to this drug in *Staph. aureus*, probably partly because it was hardly ever used, because it was so toxic.[133] It was called Mississippi Mud, because of the impurities present and it required a further purification process before most people would be prepared to use it, but with the emergence of MRSA, vancomycin began to be used much more widely.

Teicoplanin, the other glycopeptide, is much less toxic and serum assays were not made available [by the manufacturer to doctors] for it to emphasize its safety. Ironically, because many doctors were frightened of vancomycin toxicity, they were also wary of teicoplanin, and tended to underdose because they could not perform assays. As a result, the early literature has many reports of failures with teicoplanin and this may be part of the reason why teicoplanin was never licensed in the US. Although teicoplanin works very well, it is not used nearly so much as vancomycin.[134]

In the mid-1980s vancomycin usage increased dramatically for the treatment of MRSA, methicillin-resistant coagulase-negative staphylococci and as an oral agent for *Clostridium difficile* diarrhoea.[135] Inevitably, about the same time reports

---

[133] Professor Harold Lambert suggested in the 1998 Witness Seminar that vancomycin was, at that time, the accepted reserve drug [Tansey and Reynolds (eds) (2000): 54]. For other antibiotic policies of the time, see Williams *et al.* (1960): 218.

[134] Professor Gary French wrote: 'Inevitably, around the same time reports began to appear of vancomycin resistance in enterococci [Woodford *et al.* (1995)]. This was high level resistance to both vancomycin and teicoplanin – hence glycopeptide-resistant enterococci, (GRE), also called vancomycin-resistant enterococci (VRE), which is encoded by a transposon on a transferable plasmid. GRE have since become more common, especially in the US. In Europe, GRE are found in the bowels of animals fed the glycopeptide avoparcin as a food supplement and in faeces of normal people in the community [Bates (1997)]. The evidence is incomplete, but it is widely accepted that the use of avoparcin in animal husbandry is associated with the emergence of GRE in animal faeces that may enter the food chain and colonize humans [Witte (2000)]. There has been continuing concern about the possible transfer of glycopeptide resistance from GRE to MRSA. Transfer to *S. aureus* was first done experimentally on the skin of nude mice by Bill Noble at St Thomas' Hospital [Noble *et al.* (1992)]. However, the first report of a clinical isolate of fully vancomycin-resistant MRSA did not occur until ten years later [Sievert *et al.* (2002)] and there have been only a handful of such isolates since then, all from the US. Thus, transfer of high level vancomycin resistance from GRE to MRSA is fortunately a rare phenomenon, despite the millions of opportunities there must have been for these organisms to interact in hospitalized patients.' Note on draft transcript, 4 December 2007.

[135] Kirst *et al.* (1998).

began to appear of vancomycin resistance in enterococci.[136] This was high-level vancomycin resistance on a transmissible plasmid and we were very concerned about the transfer of this plasmid into MRSA. There must have been millions of opportunities for the vancomycin-resistant enterococci to interact with MRSA, but vancomycin resistance transfer has hardly ever happened.[137] Bill Noble in St Thomas' Hospital first did it in the lab on the skin of naked mice. There's some debate about whether he should have done this, but he made a transfer in the lab, some ten years before it actually happened clinically.[138]

As you know, around the 1970s many pharmaceutical companies stopped pursuing antimicrobial therapy. It can take ten or 15 years to get a drug on the market and by then the organisms might well have changed entirely. Already by the 1970s there were something like 13 different classes of antibiotics that are naturally active against MSSA, including the excellent flucloxacillin. There was therefore no clinical or commercial pressure to develop new agents for Gram-positive infections. However, by the late 1980s, multi-drug-resistant MRSA, enterococci and pneumococci had unexpectedly emerged and treatment was dramatically compromised. No agents for Gram-positive infections were in development, but by chance DuPont had discovered the oxazolidinones in the 1970s.

These are entirely artificial compounds that had in-vitro activity against Gram-positive bacteria, but were not developed for human use because of serious animal toxicity.[139]

The Upjohn Company revived oxazolidinone research in the 1990s and discovered new derivatives that retained good antibacterial activity but without animal toxicity.[140] The first of these to be developed for clinical use was linezolid which is active against most clinically important Gram-positive bacteria and is now one of the standard treatments for serious MRSA infections.[141] The oxazolidonones are available orally as well as parenterally and are the first new antibacterial class to be introduced into clinical use for 30 years. In one of the

---

[136] Reviewed by Woodford *et al.* (1995).

[137] See note 138.

[138] Noble *et al.* (1992).

[139] Slee *et al.* (1987).

[140] Ford *et al.* (1996).

[141] Gemmell *et al.* (2006).

labs in the Upjohn research centre at the Kalamazoo was a small group of people headed by a scientist called Chuck Ford. They investigated these compounds before it was clear that they were needed, because they were inherently interesting. Every year he went to the Board and said something like: 'We need more money for this interesting research', and they were given it without much need to prove commercial usefulness. Eventually they emerged with linezolid, the first member of a new class of antimicrobials for 30 years. Meanwhile, just at the time the MRSA crises were recognized.

Nevertheless, the problem outlined by Norman [Simmons] remains: it is so expensive, so difficult and so commercially dangerous to produce new antibiotics, that a continuing supply of effective agents is unlikely. As we generate more and more resistant strains, we run into trouble because the pharmaceutical companies may not come to rescue us. That's just another reason not to generate new resistances and to try to bring them under control.[142]

Gemmell: I would just like to add a bit to the story about Upjohn. In 1982 I was sent two phials of powder, both Upjohn names, but one was linezolid and one was eperezolid, and, in fact, because of superior in-vitro activity the company went with linezolid. But it was 1982 when the first white powder studies were done.

Selkon: I forgot to complete the story a moment ago, in saying that despite the reassurance from Edward Lowbury, I was rather worried at the potential of not treating *Staph.* colonization in our burns unit and we looked for them very carefully. Whenever a patient was pyrexial, we did grow streps, we did grow *Klebsiella,* we did grow *Pseudomonas,* but we have never ever had a *Staph.* septicaemia. I would very much like to know if anybody dealing with burns has in fact had a *Staph.* septicaemia.

Sutherland: In reply to Dr Simmon's plea for accelerated testing of antibiotics, I have great sympathy, but if I recall right, the early penicillins were in the pre-thalidomide theory period and toxicity was not really appreciated. I must say, when I look back at Brockham Park in the early 1960s, we were testing something like two new penicillins a week in human volunteers. Members of staff were paid three guineas [£3.3s.0d; £3.15], plus a hearty breakfast, to

---

[142] Dr Geoff Scott wrote: 'Several successful studies to control resistant organisms by limiting antibiotics were published in the early days. For example, Mary Barber and her colleagues (1960) showed that penicillin resistance in staphylococci could be reduced by restricting penicillins.' Note on draft transcript, 14 January 2008. See also Appendix 2, page 82.

give five or six samples of blood – no follow-up, no group medical attention afterwards. Penicillin is a very safe molecule.[143] I am not sure I would suggest that for other antibiotic molecules.

Stewart: I wonder in this respect if the methodology has some gaps. I would be interested to know what some of the experts like Dr Sutherland would say about this. If you take the usual method of testing, it is gravimetric, it's micrograms per ml [µg/ml]. This means that you can spot differences in molecular weight. If you remember the sequence of BRL syntheses, methicillin was number 1241. There was a number before that called 1060 (or was it 1061?), in which I had taken an interest because of the known activity of an additional amino group in the side chain. This didn't, at first, appear to hold too much promise when assayed by the gravimetric method. But, when assayed chromatographically, it turned out that there were two epimers: one was highly active, that was the D(-) epimer, which is what you would expect in something which is an oligopeptide which penicillins are, and the L(+) was not active by comparison. The reason that Beecham's very wisely went to ampicillin was by a differential calculation of the amount of activity that was contributed by the active D(-) epimer.[144]

Now a similar thing has occurred to me with regard to vancomycin, and some of the others like daptomycin, in that category. Vancomycin has a molecular weight of about 1800, penicillin G has a molecular weight of about 220, methicillin has a molecular weight of about 230 or 240, and so on, and ampicillin similar. Now, if you then assay by the µg/ml, do a micrograms per molar comparison, and then compare that with the gravimetric one, then you find, of course, that the penicillin molecule – like the ampicillin molecule, like the methicillin molecule – is very much more active than the vancomycin molecule, or the daptomycin molecule, or some others that are much heavier. So in terms of the assay, the number of molecules which are active could appear to be too small to be any use in the assay. You could reject that in terms of toxicity. So you need a differential

---

[143] Professor Gordon Stewart wrote: 'This fact is often forgotten because of fear of allergy, which was largely overcome by cessation of topical therapy and removal of macromolecular residues of manufacture or storage of natural penicillin and ampicillin that have allergenic and therapeutic activity in the nanometric range, but are safe at gramme per day levels though if given as potassium salts in high doses, toxicity can be caused by the medication. Blood levels, counts and renal function should be tested when antimicrobials are used in high or continuous dosage, to facilitate and regulate high dose therapy for "resistant" infections. Used this way, isoxazolyl penicillins could be a safer option than vancomycin in treating refractory MRSA or other amphoteric, high molecular weight antigenic glycopeptides.' Note on draft transcript, 8 December 2007.

[144] See note 38. Ampicillin was known as BRL1341.

which will be calculated in terms of the molar strength. Would you have any observation from that, Robert [Sutherland][**From the floor:** I would need notice of that question]. Well, it was your company that I had the arguments with.

Richmond: From my experience in the pharmaceutical industry fairly recently, most people measure things in terms of nanomols, so I don't know that that argument holds up.

Can I make a comment about the pharmaceutical industry and its search for antibacterials? It seems to me an interesting change has come over the world; it's something that I have been quite heavily involved with for the last two or three years – or more, five years. The major companies have actually abandoned their work in anti-infectives. I suppose the most spectacular was Roche who decided that they wouldn't pursue a follower for Rocefin, when it was making them well over $1 billion a year. But what they have done, of course, is to fund a very large number of small start-up companies, particularly in the US, but a few in this country, with a view to using the originality and excitement of the people that run those companies to find new molecules, at which point they will buy them. I think we are seeing part of the evolution of 'big Pharma', which is the setting up of their out-of-house research activities, but under control, and it's been particularly noticeable in the area of anti-infectives.

Stewart: Is this what you would call 'in-sourcing' or 'out-sourcing'?

Richmond: Well, they have put in the money and it's risk capital.

Newsom: Can I just go back to vancomycin for a minute? I have the original vancomycin data sheet, which tells me that it is a substance which is 82 per cent pure, and vancomycin I think was licensed by the FDA, it must have been in desperation, within 18 months of the fungus first being isolated. So that was a 'needs must' situation.

Stewart: And justifiably, too.

Newsom: Yes, well, they thought so.

Stewart: Yes, it is a product of a different genus and a different line of thinking, and there is room for exploration there. One of the difficulties is that opportunities for patents with molecules are limited and are explored very widely for other reasons, for example the hydroxyquinolines and the quinoline group generally. This is what I think obstructed quinacillin, which was an obvious lead to follow. There is now some evidence that new forms of quinolines might be more active.

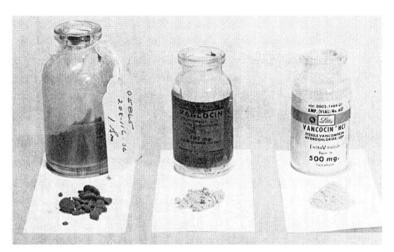

Figure 11: Stages of vancomycin development, 1956–81. L: 'Mississippi mud', early sample of compound 05865; M: vancomycin purified with picric acid precipitation; R: vancomycin hydrochloride. Griffith (1981): S202.

Gould: In terms of fast tracking new agents, I think the US FDA has done that recently for anti-retroviral drugs. But until two or three years ago, probably, it was highly arguable whether there was a perceived need. There was great debate, and still is in some quarters, about the relative pathogenicity of MRSA *vs* MSSA.[145] While two or three meta-analyses definitely now ascribe at least double the mortality to serious MRSA infection, compared with MSSA, there are certainly going to be strain differences and complications to do with the underlying risk factors in patients. But, I think, recent evidence does confirm that even with rapid institution of appropriate therapy against MRSA infections, the outcome still is inferior. I think Gary's point about flucloxacillin and oxacillin really being superb drugs that have not yet been equalled for MRSA, with the possible exception of linezolid, the jury is still out. We have to do a lot more public relations with pharmaceutical companies, if we are really to convince them of the need for new drugs.

Stewart: Well, this is why the National Research Development Corporation (NRDC) was formed, and in its day it was very successful. I don't think we would have had cephalosporins without the NRDC.[146]

---

[145] Gould (1958); Anon (1958b). See also Whitby *et al.* (2001); Salgado *et al.* (2003); Shorr (2007).

[146] For the background of the NRDC's worldwide patenting and licensing scheme and the royalties generated by the cephalosporins, see the papers and correspondence of Sir Edward Penley Abraham (1913–99), GB 0161, held at the Bodleian Library, Oxford; Keith (1981). See also Tansey and Reynolds (eds) (2000): 38–43.

Hamilton-Miller: I know for a fact that Schering dropped everninomycin, which was a very good anti-MRSA drug, because they didn't think there was enough market for it.[147] I am sure this sort of thing also happened at other drug companies. Concerning the oxazolidinones: DuPont were originally working on agrochemicals, with no interest in medicine. By chance, they picked a useless agrochemical compound off the shelf, found it was an active antibacterial and developed the series.[148]

Bud: Shall we move on then towards another section and I think we have persuaded John West to say a few words to kick off this section on changing attitudes to infection.

Professor John West: I was asked to say something about antibiotics and surgery, and first I should say that I think I am here under false pretences. I am a clinical physiologist with an interest in the history of medicine, but I don't know much about MRSA. But perhaps I can tell you an anecdote about my father, an orthopaedic surgeon trained in Liverpool in about 1935, and then returned to Adelaide, Australia, where we come from. (Incidentally, I just might mention that Howard Florey also came from Adelaide, because we have already heard about Fleming and Chain.) When my father got back to Adelaide with his orthopaedic degree, he had a very frosty reception from the general surgeons, who claimed that they could do anything that an orthopaedic surgeon could. But then two important things happened: one was the outbreak of the Second World War. My father was sent to a hospital in Palestine, where he looked after a lot of Australians who had been fighting in North Africa, and had lots of terrible bone injuries. The other important thing was that penicillin came along. Penicillin, according to my father, absolutely made all the difference to orthopaedic surgery. There was no way you could do things to bones that orthopaedic surgeons wanted to do without penicillin. It completely revolutionized that area, and I presume it is still an important factor. Incidentally it is interesting that although the general surgeons, as I said were rather arrogant in the late 1930s, it turns out now – at least in our institution – that the orthopaedic surgeons are the top dogs, and the general surgeons are actually relatively minor. I think that as far as orthopaedic surgery is concerned, the introduction of antibiotics was absolutely critical.[149]

---

[147] Dr Geoff Scott wrote: 'Everninomycin caused reversible renal failure in about 1:15 volunteers, which was one major reason the development was discontinued.' Note on draft transcript, 14 January 2008.

[148] See Daly *et al.* (1988); Gregory *et al.* (1990).

[149] See Reynolds and Tansey (eds) (2007a).

Bud: Do people want to talk about the attitude of surgeons when they started encountering antibiotic-resistant bacteria, and cleanliness?

From the audience: Ogston became famous for operating on knock-knees. He had to change from being an ophthalmic surgeon into an orthopaedic surgeon and he later became Surgeon to the Queen. So, there you are, it's very important, they all went on.[150]

Shanson: My experience during the last 30 years in many hospitals suggests that most surgeons generally didn't think much of MRSA. Patients were colonized, or often only had mild wound sepsis, and the surgeons weren't very impressed when you asked them to isolate their patients. I once had a difficult urologist problem. During one week at St Stephen's Hospital [now part of the Chelsea and Westminster Hospital, London], several urological patients developed deep-wound sepsis, within a short time of operation.[151] There was another hospital in north London where this urological surgeon worked and the microbiologist phoned me and said have you got any problem with Mr X's patients, because this hospital also had a lot of deep-wound sepsis in his patients, and that was within a fortnight of our experience.[152] The urological surgeon didn't think much of MRSA, he didn't think he was doing much harm, but I had to investigate him and stop him operating. He applied various antiseptics to his skin before I swabbed him, but we none the less isolated the epidemic MRSA strain from him. He was initially resistant to treatment and had an eczematous skin lesion on his hand. We had to change his dermatologist to get his skin better and after various treatments he became MRSA-negative and started to operate again. I got some advice from Graham Ayliffe, at the Hospital Infection Research Laboratory in Birmingham, on various impermeable gowns for him to use. Unfortunately about eight months later, the same MRSA strain came back again with a new outbreak of theatre-acquired wound sepsis involving only his patients. This time, he was afraid for his livelihood, and I had to get the 'three wise men' of the hospital and the house governor to stop him operating again. Eventually we got his cooperation, and finally managed to clear him permanently of MRSA carriage.

Phillips: Another anecdote about surgeons. My experience over the years is that they respond to a dramatic event. This particular dramatic event was our first couple of EMRSAs in cardiac patients. I had been giving a talk in Vienna,

---

[150] See also Newsom (2004).

[151] Shanson and McSwiggan (1980).

[152] See Professor Michael Emmerson's description on page 49.

saying that we didn't have MRSA and I came back to find two patients in adjacent beds, one of whom had an aortic aneurysm infected with MRSA. The old tradition was that you put sawdust around these unfortunate people, for fairly obvious reasons. His aorta duly ruptured into his oesophagus. The surgeons then believed and supported us. And, they spread the word.

Dr Geoffrey Scott: I am sorry I have remained curiously silent for so long, but I just wanted to reiterate what Stephanie had said a bit earlier on that this is an extraordinary meeting. It's a great pleasure to be here and listen to everybody's witness to what happened. I read that in the 1950s when penicillin-resistant *Staphylococcus aureus* emerged that there were reports of outbreaks of surgical sepsis of about 25 per cent in many hospitals, to the extent that the MRC started to look into this.[153] They went around the hospitals, and of course they found that there were very high rates of post-surgical infection with penicillin-resistant *Staph. aureus* in all of the hospitals. They went to the surgeons and asked, 'How many infections do you think you have?' and they all said, 'Nil'. And, as I teach the medical students, the reason for that is that they liked to examine patients with lumps and bumps and operate on them, but they are not really terribly interested in outcome and left post-operative care to their juniors. Of course, if they look inwardly at their outcomes, then perhaps reflecting on patients who had had a bad outcome would make life quite difficult to deal with. After all, they are the people who are actually going to do things to patients and for that they have my highest respect.

Nowadays, surgeons, of course, do the most extraordinary things to patients in terms of the risks to them and to older and older patients as well. People become more and more susceptible to infection. Surgeons work in a sea of multi-resistant organisms, not only the Gram-positives, but also the Gram-negatives, some of which are very resistant. And yet, even in a hospital like UCH next door, given that there's a certain amount of MRSA floating around, moving around, if you go to any one individual surgeon and say, 'Do you feel that this is a problem?', or to a physician within, say, a medical committee, they don't perceive it to be a problem. This is because if you take any individual person, they will, perhaps, see one infected patient every three months. If you have a sufficient number of consultants, the amount of work is divided up rather in such a way that the number of cases is not so great as to have a big influence on their brains. That is difficult for infection control people, because they are seeing all the cases every day. I tend to have a very jaundiced view of the outcome of work within our

---

[153] MRC, Cross Infection Committee (1951). For an earlier MRC report, see Hill (1934).

hospital, because I see the patients for whom the treatment has gone wrong, who have problems, who have infections and so on. And it was ever thus. So, I think the attitudes of the people within this room might be rather subtly different.

Duckworth: I concur with what Ian Phillips said about a dramatic event having an incredible capacity to focus the mind. In the early days of the London Hospital outbreak, it affected cardiothoracic surgery – chest walls just fell apart post-operatively, so that you saw everything you shouldn't see inside the chest. That certainly got our surgical colleagues on board pretty fast. Turning to the recent surgical site infection data, many hospitals participate routinely in surgical site infection surveillance, and some of it is mandatory, for instance in orthopaedics.[154] Now the overall infection figures are very worrying from the point of view of MRSA, but, as you would expect, *Staph. aureus* accounts for the majority of these post-operative infections, but around 60 per cent of the *Staph. aureus* infections are due to MRSA.

Simmons: I retired in about the mid-1990s. I had one surgeon who had TB and didn't want to stop operating, but we persuaded him in a few short sentences. I never had a problem with any other surgeon, physician, administrator, manager of any sort, not doing what I thought was necessary to end an outbreak. Never a one. If I said a ward had to close, it closed, and we told him [the surgeon] it was shut. If I said the patients needed isolating, they were isolated. I never deferred to the management in terms of the prevention of infection. That's my job. I have a responsibility to patients, that's what I regarded it as. But I never for one minute had a surgeon who refused to do what I told them to do in respect of infection. If he wanted to operate, I never interfered with his operating technique, and he wouldn't interfere with my responsibility.

French: I agree with two or three points which have been made about surgeons paying attention to infection. There is now evidence from meta-analyses to show that patients with MRSA infections do worse than those with MSSA, when other factors are controlled for.[155] This is certainly true anecdotally, and our surgeons – especially those in cardiac, vascular and orthopaedic (implant) surgery – are now frightened of MRSA infection. I think this is the first time surgeons have been scared of bacteria for 40 years.

---

[154] Health Protection Agency (2007c).

[155] Whitby *et al.* (2001); Cosgrove *et al.* (2003); Engemann *et al.* (2003).

Surgeons are also very concerned about litigation. At the present time MRSA is nearly always acquired in hospital, and the lawyers know this. The surgeons quite rightly are very concerned about the legal aspects. A patient who comes into hospital for a routine operation, acquires MRSA and suffers serious post-operative sepsis, is likely to take the hospital and surgeon to court for negligence. A similar infection with MSSA may not be challenged because MSSA is more likely to be acquired in the community. MRSA are regarded as being hospital-acquired, and therefore preventable.

Chattopadhyay: I must be the odd man out to be defending the surgeons. I must say that the culture is changing, and it is changing fast. There are several reasons for it, we have already heard from Gary [French] that they are under the spotlight. First, their post-operative infection rate; second, their post-operative mortality; and third, post-operative adverse events, like taking patients back to the theatre again; and finally, re-admission. Surgeons are very serious about these issues and I understand that the league table is there, for cardiac surgeons, for everyone to see, and is in the public domain. But what we also found very helpful was to have an active surgeon, Mr Ravi Kunzru, on our [Whipps Cross] Hospital Infections Control Committee.[156] In this case it was an orthopaedic surgeon who was really brilliant, and he used to take these bad messages back to his colleagues, and we got nothing but support from this orthopaedic surgeon and I think unless we can take them on board, we are unlikely to achieve what we are aiming for.

Stewart: With certain reservations, I would also like to throw in a favourable word for surgeons. For one thing, and I mention one name in particular, Mr Harold H Nixon, a surgeon (unfortunately, now dead) of Great Ormond Street, London, who also had beds at Queen Mary's Hospital for Children [Carshalton].[157] He was the first surgeon who saw the need – I was working with him at the time – to use the new penicillins on surgical cases, and the first convincing case was one of a child who had undergone neurosurgical procedures and others followed later.[158] This was what first convinced people clinically that there was a strong case for using methicillin. The surgeons had been dragging their feet until then. Now that was important. But, if you think back again about the surgeons, when Ogston discovered *Staph. aureus, c.* 1865, there was no evidence that other

---

[156] See Reynolds and Tansey (eds) (2007a).

[157] Nixon and O'Donnell (1961).

[158] Stewart *et al.* (1960); Callaghan *et al.* (1961).

surgeons paid a lot of attention to this, except that he himself was a surgeon.[159] Two years later in Glasgow Royal Infirmary, Lord Joseph Lister discovered cross-infection due to staphylococci, and used Ogston's work, and that was what came through as cross-infection in surgical wards. In surgical wards, with cases of fractures and so on, you got cross-infection and staphylococcal osteomyelitis and bacteraemia, septicaemia, whereas in burns units there were superficial infections, streptococcal infections, different things; staphylococci were in their own niche. So surgeons have had a key role in all this and I would say that today the main problem in MRSA is with the surgical side, and it's essential that surgeons are kept in the picture all the way. My experience is that if that is done, then they are immensely helpful. But there are difficulties now, because many surgeons are not in control of their own wards, they can't open them, they can't close them, sometimes they can't admit patients. Sometimes if they do admit patients, the patient is transferred without their being consulted. I have kept tabs on half a dozen such cases, very severe, and one fatal, in the past two or three years, and I have been myself consulted about that. There is a strong administrative pressure, which defeats clinical independence. This has to be recognized.

Bud: I think that leads us very nicely to the questions of epidemiology and I think Professor Spratt could lead off on this topic. Well, would we like to talk about it from a different angle then, about the Staphylococcal Reference Laboratories, because we have two people here, Dr Pitt from England and Professor Gemmell from Scotland.

Duckworth: What the general epidemiology has been showing is that rates of MRSA have risen dramatically, from less than 2 per cent of *Staph. aureus* bacteraemias in the early 1990s to around 40 per cent at the turn of the millennium.[160] However, focusing on MRSA as a proportion of all *Staph. aureus* bacteraemias hides the fact that methicillin-susceptible *Staph. aureus* (MSSA) bacteraemias have been rising as well. I think a lot of this focus on MRSA takes our eye off the ball as regards methicillin-sensitive *Staph. aureus*.

Coming back to what's been happening recently (I can really only speak about English data), the control of MRSA is extremely high profile. It is very high on the Government's agenda. Unlike the situation Joe was rueing earlier, it is now very much the Government who is driving action to reduce MRSA. I think it is quite sad that action has had to come from the Government and not from our

---

[159] Ogston (1984).

[160] Duckworth and Charlett (2005).

colleagues. Many of our colleagues got very blasé about MRSA. There was much complacency, an attitude of 'we have to live with MRSA'. The mortality associated with MRSA was not being recognized. Joe [Selkon] alluded to a recent paper from Oxford, which indicated high mortality, and if you look at the Office for National Statistics (ONS) data, they show that mention of MRSA on death certificates has increased in parallel with the rise in MRSA bacteraemias.[161] We are currently undertaking a study with ONS to examine this in more depth. But coming back to my point about the Government driving action, what this has meant is that reporting of MRSA bacteraemias became mandatory from April 2001.[162] Subsequently, reduction of MRSA bacteraemias has become one of the NHS' top 20 targets (a 50 per cent reduction by 2008). The Trust's Chief Executive has to sign off the Trust's MRSA data monthly and this data is being scrutinized closely, not only by the Department of Health, but also by the Prime Minister's Delivery Unit. If a trust looks as if it is not on course to meet its target, that information immediately flows along the performance management pathway, to the Strategic Health Authority (or monitor for Foundation Trusts). Trusts that are not improving are visited by Department of Health review teams to assess action within the Trust and develop an action plan. So, as you can see, MRSA control is extremely high profile. The impact of this so far on the data is that MRSA bacteraemias appear to have plateaued since reporting became mandatory, but this is against a background of ever-rising rates before then. So, although Ministers etc. are unhappy that further national reductions are not yet visible, in my book a plateau gives cause for optimism against the increases we were seeing year on year before. There is nothing to indicate that levels would have automatically plateaued, as the story from some other countries has been that MRSA levels can go on rising, well beyond 40 per cent of all *Staph. aureus* bacteraemias.

Bud: Do you or anybody else want to comment on what happened to transform the political response from what we saw earlier, what people have been talking about, which is an ambivalent attitude within the profession towards this very strong political response?

Scott: A lot of this was driven by high-profile political people being lobbied by ordinary people in the country, saying that their loved ones, their dear ones, had died from hospital-associated infection. I was the first Press Officer for the

---

[161] Wyllie *et al.* (2006). The ONS data is reviewed in Griffiths *et al.* (2004).

[162] Every NHS Trust in England and Wales must record and report levels of hospital-acquired infection after April 2001. Health, Department of (2000, 2001, 2004). For the most recent targets, see Department of Health (2008); Figure 12.

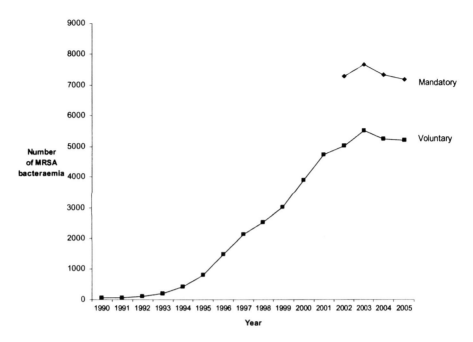

Figure 12: *Staphylococcus aureus* bacteraemia reports received under the voluntary and mandatory surveillance schemes in England, 1990–2006. Mandatory surveillance started in April 2001. Source Health Protection Agency. .

Hospital Infection Society, and at the time that we had our first international meeting, I can't quite remember when it was now – 1987, thank you, David [Shanson] – I didn't know anything about being a press officer and I said, 'What do you want me to show?' They said: 'We want you to show that there is a Hospital Infection Society'. Of course, nobody knew anything about hospital infection at that time. I would say that we had quite a successful campaign, but in fact what I learnt was that the only news is bad news. We had some very, very interesting statements like [those from] Richard Marples, our great leader of our Staphylococcal Reference Lab, saying that the only way to get rid of MRSA would be to burn the hospitals down.[163] There is one nice quote, and I think it was Peter Davey, who said that the surgeons were the reason for MRSA and all these surgical infections. We were pushing quite hard to get some recognition and the newspapers were picking up on stories all the time, but in fact it rebounded on us in a way, because when people started to get MRSA sepsis and perhaps even succumb from it, that became very serious public and political news. Of course, MPs are lobbied, and if relatives of MPs get this infection, then they go to the

[163] See, for example, Marples (1981); Marples and Richardson (1982); Richardson and Marples (1982).

Secretary of State for Health and say, 'Well, what are *you* going to do about it?' Of course, in this Government they set targets, as Georgia [Duckworth] has said, they set a target for us to reduce our MRSA. But, as Ian Phillips said, we do not know the scientific determinants for the spread of this organism. We do not know, because we have not done the proper research to find what it is that makes this organism tick, and I don't mean MRSA, I mean *Staphylococcus aureus*. Gary [French] is right to say that MRSA is probably worse for you than MSSA [methicillin-sensitive *Staph. aureus*], but that is the status today, and if we had virulent penicillin-tetracycline-resistant *Staph. aureus* floating around our hospitals – 'phage type 80/81 as in the 1970s – I could promise you that some of our patients would be doing exceedingly badly, if they picked that up.[164] There are so many linked factors in the evolution of this organism within our hospital practice, that it's almost impossible to get our brains around them. So, I would say that a target has been set, but it's rather like King Canute being asked to hold back the waves. He didn't know quite how to do it, he hadn't been given guidance on the moon, and astronomy, and actually how to change the waves and so on, and similarly, we have been set a target for MRSA rates and nobody really knows what it is that we have to do to reduce the burden of the bug in our hospitals. I think that's a very important thing for us to consider as a group.

Selkon: It is very pleasing to hear that the Prime Minister's office is considering, involved and worried about MRSA: good intentions. But until we have the great pressure removed from our hospitals to get the waiting lists down, we have got to keep on running at up to 102 per cent bed occupancy in our surgical wards. The present newly arrived efficiency expert appointed to our hospital has now decided that the operating theatres were only used for 87 per cent of the time available and this is to be increased to 93 per cent, and the number of pathology investigations should be reduced. Perhaps if we reduced the number of wound swabs, the prevalence of MRSA infection will also be reduced. Until the load on the hospitals to undertake non-essential surgery is reduced, I don't see how we can, with our inadequate isolation facilities, achieve safety for our patients.

French: I agree with Geoff [Scott] in this. When patients die with MRSA sepsis, their grieving relatives notice that the wards are crowded, dirty and understaffed and are rightly outraged. Things are changing dramatically, but until recently, standards of cleanliness and hygienic practice in some hospitals have been poor. Microbiologists did not draw a line in the sand over this because it has

---

[164] See, for example, Robinson *et al.* (2005).

been difficult scientifically to prove a clear association between environmental contamination and infection and managers took advantage of this to cut costs.

R E O Williams said that *Staph. aureus* was his favourite organism, because it had designed hospitals. All the standard hygienic practices in hospitals that we are supposed to ensure is done – smooth floors, cleaning, damp mopping, hand washing, sterile uniforms, separate operating theatres, changing bedclothes, filtering the air – everything is directed primarily at the control of *Staph. aureus*.[165] We need to continue to maintain hospital hygiene.[166]

I have to add that the microbiologists didn't help, because there was a big debate – very interesting historically – as to whether MRSA was important. Many people felt that we were wrong to take MRSA more seriously than MSSA and that a great MRSA industry has arisen that was totally unrelated to the importance of the organism in their hospitals. For some hospitals this may have been true.[167] When the 1998 Guidance on the control of MRSA came out, there was a bitter exchange of letters in the *Journal of Hospital Infection*, in which half the country's microbiologists more or less said this guidance was a waste of time, and by implication was written by London elitists, who knew nothing about

---

[165] See, for example, French *et al.* (2004).

[166] The British Broadcasting Corporation's *Panorama* investigation, 'Undercover hospital cleaner', was broadcast on Wednesday, 13 July 2005, on BBC One. Health Secretary Alan Johnson announced that by January 2008: 'Hospitals will adopt a new "bare below the elbows" dress code, i.e. short sleeves, no wrist watch, no jewellery and allied to this the avoidance of ties when carrying out clinical activity. The traditional doctors' white coat will not be allowed. The new clothing guidance will ensure good hand and wrist washing.' Press Release 2007/0269, 17 September 2007 [freely available at www.gnn.gov.uk/environment/fullDetail.a sp?ReleaseID=314953&NewsAreaID=2&NavigatedFromDepartment=True (visited 7 February 2008).

[167] Professor Gary French wrote: 'Microbiologists were also split over the significance of MRSA. There was a widely held view, which is still common amongst some, that methicillin- and multi-resistant *S. aureus* are no more important than MSSA and that all the efforts expended on MRSA control are firstly ineffective and secondly took time and resources away from other, more important pathogens. It was strongly felt by some that microbiologists were chasing a marker (methicillin resistance) that had no relevance to clinical significance and that MSSA were equally important but largely ignored [Lacey (1987); Rahman *et al.* (2000)]. This argument continues, but there is now good evidence that MRSA infections have poor outcomes and that infection control – properly and strictly applied – can reduce MRSA infection rates. In my own view, microbiologists were remiss in not pursuing MRSA more actively in the early days of the epidemic when control was more possible.' Note on draft transcript, 3 December 2007.

infection control issues.[168] There are still some microbiologists who believe that the importance of MRSA is overplayed, although I am not among them.

Microbiologists did *not* put their foot down about filthy and crowded wards. The public did, and that's what got through to the politicians. MRSA became the focus for a growing understanding that infection control in hospitals was poor.

Duckworth: I would like to add to what Geoff [Scott] and Gary [French] have been saying about what gave MRSA the high profile for Government action. I think it was more than patients beginning to make a fuss. It was also the data itself, because that rise throughout the 1990s was very dramatic, and when we saw our data in the context of rates in countries on the continent, such as the Netherlands and Scandinavia, the situation became deeply embarrassing. That came to the attention of both the Department of Health and the National Audit Office (NAO), which undertook their first study of hospital-acquired infection and that report to the Commons' Public Accounts Committee required a Government response, further raising the profile of the topic dramatically. In an unprecedented fashion, the NAO stayed on the topic for yet a second report, making a major impact.[169]

Coming back to what Joe Selkon was saying, my role isn't to defend the Department of Health. High bed occupancy must play a major part when we compare our situation with that in some countries on the continent, but despite all these difficulties, some trusts are making major inroads on their MRSA levels. I know Gary speaks about how they have made an impact in his trust. Despite all those pressures on beds, etc., some trusts have made a significant difference. What is quite interesting is that rates now appear to be coming down in acute specialist and teaching trusts, while it is in the District General Hospital (DGH)-type hospitals that levels are still rising.

Bud: One more on this and then I think we need to talk about the Staphylococcus Reference Laboratories.

Simmons: Why has the public attitude changed? It's a good story. MRSA: it trips off the tongue; so does superbug. Even Professor Ayliffe used the word.

---

[168] British Society for Antimicrobial Chemotherapy, the Hospital Infection Society and the Infection Control Nurses Association, Combined Working Party (1998).

[169] National Audit Office (2000, 2004). House of Commons, Committee of Public Accounts (2000).

It makes me want to puke; it's an awful word; what's super about it? It's an awful word for a scientific meeting, it really is. It's not super. But it trips off the tongue. Leslie Ash gets an ordinary staphylococcal infection.[170] And, I got a magazine through my letterbox yesterday, which talks about MSSA, a variant of MRSA. Rubbish.

Targets: I hate to tell you but the staphylococci aren't frightened of the Prime Minister, they don't take any notice of politicians – they don't care who's in No. 10 – or even of the Chief Executive. Let me tell you, I spoke to someone, I had better not let you identify the person, who went to Russia at the request of the PHLS a number of years ago to assist the Russians in the control of post-operative infection. The Soviet government decreed that post-operative infection would not be more than 2 per cent, and the person said to me, 'I learned of all the ways that government intervention could keep it below 2 per cent.' Every time there was a report, someone went and visited the laboratory, and said: 'Are you sure? Do you think taking all these swabs is a good thing?' They want to penalize hospitals that have a high infection rate. The way to get this down is through the application of science. I am not all that convinced that the intervention of politicians is a good thing. Finally, the bed occupancy rate in the Netherlands is 60 per cent, the occupancy in Britain is over 80 per cent. A paper has appeared recently showing a straight line correlation between occupancy and MRSA infection rate, or rather the MRSA bacteraemia rate.[171] As far as I know, there is no published figure for post-operative infection after 1996, and that was ten years ago. All we know is about MRSA bacteraemia. We don't know the post-operative infection rate now. There might be thousands. We must draw our conclusions from the bacteraemia alone. I doubt if anyone would want to find the real figure, because that might really frighten Leslie Ash. So there are lots of reasons for public change. It's a scare story and bad news which is good news if you are selling newspapers. My contact in the press said to me: 'I have looked at what you have written about this, and I am now convinced we are just going to have to put up with it.' And people will get used to it, and then it

---

[170] Actress Leslie Ash was admitted to Chelsea and Westminster Hospital, London, in April 2004 with a punctured lung and two broken ribs; she contracted MSSA during her treatment, leaving her partially paralyzed and unable to play active roles as an actress. The case against the hospital was due to be heard in April 2008, but was settled out of court in January 2008, with Chelsea and Westminster Hospital Trust admitting a breach of duty for 'shortcomings in her care', with total compensation of £5m to cover loss of earnings. Sanderson and Gibb (2008).

[171] Cunningham *et al.* (2006).

will drift out of the news for a little while and then come back. I think it's really press awareness that has driven the public concern.

Ayliffe: Briefly, I would like to say that 50 years ago our hospitals really weren't so good. They were overcrowded, with beds along the centre of the wards; they were often not clean, and staff rarely washed their hands. And a house surgeon in a general hospital came in one morning and said to the consultant: 'Welcome to Scutari'.[172]

Bud: On that note, can we move to Dr Pitt, to introduce the Staph Reference Laboratories and also to Professor Gemmell.

Dr Tyrone Pitt: I am really only here as the link with the past; I have not made any contribution of any note to the staphylococcal lexicon. The Staphylococcus Reference Laboratory was formed in 1949 by R E O Williams. I joined the laboratory in 1968, seven years after Pat Jevons had described methicillin-resistant strains.[173] The main function of the lab over the years has been to support outbreak investigation by typing of isolates from hospitals and carry out limited surveillance of strain types at local regional and national levels. Today its best contribution is added-value characterization of strains.

Originally it was a factory, a factory of 'phage typing. All isolates of public health importance were referred (and some not so) and we had to type these at a rate of 100–200 strains a day.[174] These were then read, I should stress, only by the senior staff, and often through a haze of cigarette or pipe smoke, which one just can't imagine today. I remember Liz Asheshov and Pat Jevons used to have their ashtrays at their side as they read the 'phage plates.

---

[172] Professor Graham Ayliffe wrote: 'Scutari was the base hospital in Turkey for sick and wounded troops in the Crimean war. Florence Nightingale found hygienic standards were terrible and the mortality excessively high.' E-mail to Mrs Lois Reynolds, 3 December 2007.

[173] In a letter to the *BMJ* in 1961 [Jevons (1961)]. See page 12.

[174] Dr Tyrone Pitt wrote: 'Today this has expanded to the international arena owing to the spread of successful lineages, particularly the epidemic MRSA and less so, CA-MRSA [community-acquired or community-associated multiply-resistant *Staphylococcus aureus*]. The early laboratory processed large numbers of isolates on a daily basis using standard methodology ('phage typing and antimicrobial susceptibility screen), and results reported within 48–72 hours. These were subcultured, grown in broth, agar plates seeded and phages applied with the aid of a rather Heath Robinson invention of Dr Owen Lidwell [Figure 13], a derivative of which is still used to this day. Following incubation, the plates were read for 'phage lysis patterns by senior staff.' Note on draft manuscript, 14 August 2007.

The strength of the laboratory is that it provides an overview of the range of strain lineages circulating in hospitals, their prevalence and sometimes their patterns of spread. These data inform surveillance strategies and remain pivotal in the design of interventions to curtail spread of the organisms. A common criticism of these data is that only atypical or problem isolates are often sent by hospitals, so one could get a skewed view of the significance of these strains relative to others. On the other hand, we can be assured that the laboratory receives isolates of public health importance. Over the years, the emphasis has changed from the high-throughput ethic to more tailored investigations in support of hospital microbiologists. For example, today, in addition to traditional 'phage typing, the service extends to DNA fingerprinting by pulsed-field gel electrophoresis, toxin gene profiling (enterotoxins, exfoliatins, TSST and PVL), staphylococcal cassette chromosome (SCC) *mec* type, and DNA typing by *spa* gene and multilocus sequence typing.[175]

Figure 13: Lidwell 'phage typing machine used in the Staphylococcus Reference Laboratory for almost 50 years.[176]
Lidwell (1959). Photographed at the Colindale Staphylococcus Reference Laboratory, 2007.

---

[175] See van Belkum *et al.* (2007).

[176] Dr Tyrone Pitt wrote: 'This machine is used to apply 25 phages to lawns of staphylococcal bacteria growing on an agar surface. The metal pins pick up volumes of the 'phage and deposits them on to the agar surface. The arm revolves through a methylated spirit bath and then into a gas flame which sterilizes the pins ready for the next sample. You really have to see it in action and see the result it produces; it is simplicity itself. The tube [lower left-hand corner] is a gas lighter, the tweezers are used to straighten the pins when they are hot and the gas supply is controlled by the stopcock controls [at the back of the apparatus]. They don't make them anymore and 'phage typing is rapidly disappearing into the history books.' E-mail to Mrs Lois Reynolds, 27 November 2007.

The departure from simple 'phage typing to the application of new technologies really started with the arrival of Dr Dick Marples in the mid-1970s. By the early 1980s the old guard of Jevons, Asheshov, Lidwell and, of course, Dr Tom Parker, who was the director from 1961 onwards, had already retired. Under the direction of Professor Mary Cooke in the 1980s there was an explosion in the application of new technologies, many based on electrophoretic separation of whole cell proteins and plasmid DNA. The introduction of restriction digestion of chromosomal DNA and the use of specific gene probes accelerated the definition of strain types and led to the nomenclature of the 'epidemic strains' of MRSA [see Appendix 1]. These now number 17 types [EMRSA-17], but recent genetic analysis shows that there was considerable duplication among them owing to excessive discrimination attributed to minor differences in 'phage typing patterns and genetic lineages, there are probably only five or six lineages there. Some EMRSA were not that 'epidemic' and never spread to more than two or three hospitals. However, the fact remains that today, about three-quarters of isolates are represented by two strain lineages EMRSA-15 and EMRSA-16.

At the turn of the millennium we were typing something like 50 000 isolates a year. Today, the number is under 10 000, although we have witnessed increased demand for toxin gene profiling given the public health impact of PVL-associated disease and the emergence of community-acquired MRSA and a renewed interest in MSSA. Angela Kearns, the current head of the service, is even contemplating abandoning 'phage typing, like the Americans have done, and going to molecular-based typing, which might evoke a sharp intake of breath from the spirits of Williams, Jevons, Asheshov, Marples and Parker.

I met Dr Parker last year a few months before he died.[177] I had sent him the paper by Mark Enright and others in the *Lancet*, to which Brian Spratt has referred, on the re-emergence of 'phage type 80/81 strains and he read it with interest. He commented, 'If only we had then the tools they have today, we could have learnt so much more'. I think this is no doubt a common reflection of elderly scientists reminiscing.

Gemmell: If I could just give you a Scottish perspective on Reference Laboratories. We are much younger, infants compared with our English colleagues, we only started nine years ago in 1997. In fact we started off with about 2000 strains being sent to us, but very soon we actually reached 4000, 5000, and we are

---

[177] Dr Tom Parker died in 2006. See Biographical note on page 122.

now running at 10 000 strains sent to us from laboratories serving Scottish hospitals. As our English colleagues know, I am sure, we actually stopped doing 'phage typing in 2001, so we have already given up 'phage typing as not being appropriate to our needs. We are doing all molecular tests – PFGE, PCR ribotyping, and *mecA* gene detection tests – as well as the toxin typing. But one thing we are trying to do, as well as trying to follow the epidemiology of MRSA in Scotland, and we have seen over the past nine years, this spectrum of 75 per cent EMRSA-15s to 20–25 per cent EMRSA-16s and they are really not changing anything. But certainly what we are starting to do is something called snapshots, taking a short period of time in one particular hospital laboratory, and looking at all the MRSA strains, to try to see whether there are any trends year-on-year in the types of staphylococci [MRSA] that they are detecting. That's where we are in Scotland.

Bud: I am aware that the evening is going on and the drinks are approaching, but I want to ask Bernard Dixon to continue a theme which we discussed earlier, about the public response in the 1990s. I think a lot of what we were talking about – the press response and the public, and the Government response – leads up to what Bernard might say.

Dr Bernard Dixon: I am aware that I am now all that stands between you and the drinks that have been mentioned. It helps me enormously that I have no data, unlike every other person who has spoken. I have been asked to talk about public attitudes. If you want to know about public attitudes to, say, GM food or animal experimentation, there are places you can go, for example, to the *Eurobarometer*, which is done regularly in all the EU countries, and it will tell you what people know about that subject, and what they think about it.[178] There are, as far as I know, no comparable data regarding antibiotic resistance. Maybe it would be a good idea if questions on that were included in the future *Eurobarometer* studies.

Very briefly, what I can do is the next best thing, which is to talk about the media. There is a link with the media, which reflects the public at large, their views and their interests. The media also on occasions lead public opinion,

---

[178] The *Eurobarometer* study is sponsored by the European Commission in Brussels and the UK coordinator is Professor George Gaskell, London School of Economics. See the International Data Service (ESDS International) website, part of the wider Economic and Social Data Service (ESDS) and a national data service that came into operation in January 2003, jointly funded by the Economic and Social Research Council (ESRC) and the Joint Information Systems Committee (JISC), at www.esds.ac.uk/international/support/I33089.asp (visited 10 December 2007).

influence it certainly. I wanted to say a few things about that and give one brief story by way of analogy with the MRSA story. It is quite surprising that there has been comparatively little said about *Staph. aureus* resistance and more recently MRSA in the media until the past ten years. But, of course, over that period the problem itself has grown and there have been a number of occasions when the media have reflected that. The 1994 POST (Parliamentary Office of Science and Technology) Report, although primarily about TB, did get a lot of coverage on the general subject of drug resistance. In January 1996 the BBC's *Panorama* did a very good programme on drug resistance, which was linked with the emergence of vancomycin resistance in enterococci.[179] Then of course there was the House of Lords Select Committee in 1998, which was the lead item on at least three channels and, I think, possibly all of the BBC morning news bulletins.[180] This clearly struck a chord with the gatekeepers in the media, the editors of programmes of that sort. Since then the subject has been pretty regularly in the news. The Government response to the Lords *Report*, the Advisory Committee being established and other points were all followed very closely by the media.

I must just add one point in defence of the slight criticism of the media that I heard earlier. I have seen very little sensationalism, very little exaggeration, in the media on this problem. I have two things here that don't come from a tabloid newspaper. The first is from *Nature,* an article by Julian Davies, and the title is 'Bacteria on the rampage'. The other is from the *Lancet,* the world's major medical journal, on the occasion when vancomycin-resistant *Staph. aureus* first emerged and the title is 'Apocalypse Now!' These colourful headlines come from learned journals, not from the [popular] media. Let me go back to my chronology. There have been certainly lots of local stories, based on people with particular experiences which they took to their local newspapers, some of which got into the national newspapers. That, I think, has been an increasing trend. But then, of course, there was last year's General Election [2005], when the whole question of MRSA became a political football between the two main parties. Michael Howard, whose mother-in-law's death was apparently linked with MRSA, made a very strong attack on Labour's record in hospital cleanliness.[181] This was responded

---

[179] 'Superbugs', the BBC TV *Panorama* programme, was broadcast in 1996. A 40-minute videocassette is held as 771V in the Moving image and sound collections, Wellcome Library, London.

[180] House of Lords, Select Committee on Science and Technology (1998).

[181] See, for example, a BBC news item on Friday, 29 April 2005, at http://news.bbc.co.uk/1/hi/uk_politics/vote_2005/wales/4497675.stm (visited 12 December 2007).

to by John Reid, the Health Secretary, and he, of course, blamed the previous Tory government for not combating MRSA.[182] Even more interesting than these exchanges was the great efflorescence of letters in the national newspapers, both the serious broadsheet newspapers like *The Times*, and also the tabloids – they all had letters from people about the problem, including experts.[183] This was interesting because it was something of a cacophony, which I would have thought indicated to the average intelligent reader that there was no settled view as to what was the main problem, or certainly what was the most likely solution. There were letters about the organism itself, how it behaved, there were letters about the privatization of hospital cleaning services. There were articles and letters about the Netherlands and their policy of screening and isolating patients, and lots in the way of different ideas and suggestions, even from people who appeared to be experts. The average interested reader could simply conclude that there was no single view about this problem and certainly about the priorities for action, what needed to be done. Finally, I want to say one thing about a parallel case, which I think illustrates what hasn't happened with MRSA in terms of influencing public attitudes. The example I have in mind is E S Anderson [Andy], whom Mark Richmond mentioned earlier.[184] I want to contrast what he did way back in the late 1960s and early 1970s, with what has not happened with MRSA. Andy was one of the people who did a lot of early work on *Salmonella typhimurium* in particular, strains that were multiply resistant and he used 'phage typing to trace the movement of these organisms from farm animals into humans. He was very concerned at what he saw as the abuse of antibiotics in animal husbandry posing a serious problem for human health. Looking back, and it may be that there are people in this room who share this view, the contribution that the agricultural misuse of antibiotics has made may not be as great as Andy believed it to be. None the less, it's still interesting that he made it his business not only to write papers in *Nature* and elsewhere, but to lobby much more widely in order to influence public opinion and political action. He spoke to me as editor of the *New Scientist*, and we published quite a lot of articles by him and about his work, as well as editorials.[185] He also talked to a man called Anthony (Phil) Tucker, who

---

[182] For details of Government guidance on healthcare-acquired infections, see *Hansard* (House of Commons, 22 March 2005), cols 751–4.

[183] See, for example, Naughton (2005). See also BMA summary of issues March-April 2005 at www.bma. org.uk/ap.nsf/Content/hpd24marto15apr05#1Carvell (visited 6 March 2008).

[184] See page 35.

[185] See, for example, Anderson (1974a and b).

was then the Science Editor of the *Guardian* newspaper, which devoted a lot of coverage to this whole question of the agricultural uses of antibiotics.[186] All of this led to the setting up of the Swann Committee and restrictions that were placed on the inclusion of tetracyclines and penicillins in animal feed stuffs for growth promotion and prophylactic purposes.[187] Now whatever you think about that, at this distance in time, it's still interesting that Andy decided to give a substantial amount of his time and effort to lobbying in that sort of way. It was successful; it did result in government action. I now wonder whether there is, indeed, a settled view in the community of people concerned about MRSA as to how to combat the problem? Is there really a settled view upon which really vigorous lobbying could be conducted? I am not sure.

Stewart: I think we are winding up now, and in so doing I would like to follow through what Dr Dixon has said: credit where credit is due. This subject has been in need of critical attention for several years, especially since 1998, when the problem enlarged everywhere and especially in this country. In that year, the Wellcome Trust had a Witness Seminar on post-penicillin antibiotics, some of you were there.[188] At that time it became obvious that problems like this needed more attention. Some of us did feel that action should be taken and the Wellcome Trust was more than willing to sponsor that. But these things don't just happen easily, it's much more difficult and I think this is where we must acknowledge the unique role of Robert Bud. Without his support and pressure, this would not have happened. What we have today is a free exchange of opinion by witnesses, by informed people who all know what they are talking about. That is comparatively rare nowadays. So I think we should end with a vote of thanks, not only to Robert and the Wellcome Trust, but to the History of Twentieth Century Medicine Group who have helped in this way, and we are very grateful to them.[189]

---

[186] For example, see Tucker (1978, 1986). Anthony Tucker retired from the *Guardian* in 1988, died in 1998 and his obituaries of well-known scientists continue to appear, such as Tucker (2006).

[187] Agricultural Research Council and Medical Research Council, Joint Committee on Antibiotics in Animal Feeding (1969). The Committee was chaired by Professor Sir Michael Swann FRS (Lord Swann from 1981). See also Datta (1969).

[188] Tansey and Reynolds (eds) (2000).

[189] Professor Gordon Stewart provided additional material in January 2007, which will be deposited along with the tapes and other records of the meeting in GC/253 in Archives and Manuscripts of the Wellcome Library, London.

Tansey: On behalf of the Wellcome Trust Centre, may I add our thanks to all of you for attending this meeting. The fact that we have gone over time and most people have remained is a very strong indicator of how important this has been and it has been a very great pleasure and privilege to listen to all your reminiscences and your comments. I would particularly like to thank Daphne, Wendy and Lois, my staff who organize and run these meetings.

Appendix 1

Characteristics of UK epidemic MRSA, 2007[190]

Epidemic MRSA	'phage pattern[b]	Antibiotic resistance pattern[c]	Toxin genes[d]
EMRSA-1[a]	(84)/85/88A/(90)/932	P T E (K) (G) S	A
EMRSA-2	80/85/90/932	P E	A
EMRSA-3[a]	75/83A/(83C)/932	P E (K) (G) (N) (Cip) (S)	-
EMRSA-4	85/(90)/932	P T E S	A
EMRSA-5	29/6/42E/47/54/75/77/84/85/81	P T K G Rif	A,B,C
EMRSA-6	90/932wk	P T E K N Ba S	A
EMRSA-7	85/932	P T E S	A,C
EMRSA-8	83A/83C/932	P T S	-
EMRSA-9	77/84/932	P T E K G S	-
EMRSA-10	29/75/77/83A/85	P T E K G	A,B
EMRSA-11	83A/84/85	P T E K G N Ba S	A
EMRSA-12	75/83A/83C/932	P T E K N F S	-
EMRSA-13	29/6/42E/47/54	P T K G N Ba F S	-
EMRSA-14	29/6/47/54	P T K N F S	-
EMRSA-15[a]	75wk	P (E) Cip	C
EMRSA-16[a]	29/52/75/77/83A/83C	P E (K) (G) (N) Cip	A, TSST-1
EMRSA-17[a]	29/77/932	P T (E) Rif F K G (N) S Tp Cip (Mup)	A

[a] Currently circulating UK Epidemic MRSA strains.

[b] 'Phage patterns at 100 x RTD (routine test dilution); wk = weak reaction; ( ) = variable reaction.

[c] Antibiotic resistance pattern:
Ba, bacitracin; Cip, ciprofloxacin; E, erythromycin; F, fusidic acid; G, gentamicin; K, kanamycin (or tobramycin); Mup, mupirocin; N, neomycin; P, penicillin; Rif, rifampicin; S, streptomycin; T, tetracycline; Tp, teicoplanin borderline MIC [minimum inhibitory concentration] 4 - 8 mg/L; ( ), variable susceptibility among isolates. Neomycin resistance is difficult to detect by disc diffusion testing with some isolates. With gentamicin-susceptible isolates, tobramycin or kanamycin are more reliable indicators of the aadD gene responsible for neomycin resistance.

[d] Toxin genes: A,B,C Enterotoxins; TSST-1, Toxic Shock Syndrome Toxin-1.

---

[190] Health Protection Agency, Evaluations and Standards Laboratory, Standards Unit (2007b).

Appendix 2

The bacteriophage groups of *Staphylococcus aureus*, after Parker (1962)

Bacteriophage group	Strains lysed by one or more 'phages										
I	29	52	52A	79	80	81[†]					
II	3A	3B	3C	55	71						
III	6	7	42E	47	53	54	75	53	54	75	77	83A
IV	42D										
Miscellaneous	(a)	(b)									

† Strains lysed only by 'phage 81 are placed in 'phage group I.

(a) Strains lysed by 'phages belonging to two or more lytic groups.

(b) Strains lysed by 'phage 187.

Source: Williams *et al.* (1966): 25.

References

Abraham E P. (1983) Ernst Boris Chain. *Biographical Memoirs of Fellows of the Royal Society* **29**: 43–91.

Abraham E P, Chain E B. (1940) An enzyme from bacteria able to destroy penicillin. *Nature* **146**: 837.

Abramson M A, Sexton D J. (1999) Nosocomial methicillin-resistant and methicillin-susceptible *Staphylococcus aureus* primary bacteraemias at what costs? *Infection Control and Hospital Epidemiology* **20**: 408–11.

Agricultural Research Council and Medical Research Council, Joint Committee on Antibiotics in Animal Feeding. (1969) *Report on the Use of Antibiotics in Animal Husbandry and Veterinary Medicine.* Cmnd 4190. London: HMSO.

Anderson E S. (1966) Letter: possible importance of transfer factors in bacterial evolution. *Nature* **209**: 637–8.

Anderson E S. (1968) The ecology of transferable drug resistance in the enterobacteria. *Annual Review of Microbiology* **22**: 131–80.

Anderson E S. (1974a) Nalidixic acid or if you can't beat 'em... *New Scientist* (21 March) **61**: 750–1.

Anderson E S. (1974b) How not to use data. *New Scientist* (31 October) **64**: 314–15.

Anderson E S, Armstrong J A, Niven J S F. (1959) Fluorescence microscopy: observation of virus growth with aminoacridines. In Isaacs A, Lacey B W. (eds) *Virus Growth and Variation*: Ninth symposium of the Society for General Microbiology held at the Sentate House, University of London, April 1959. Cambridge: Cambridge University Press, 224–55.

Annear D I. (1968) The effect of temperature on resistance of Staphylococcus aureus to methicillin and some other antibiotics. *Medical Journal of Australia* **1**: 444–6.

Anon. (1958a) The practical aspects of formaldehyde fumigation. Monthly *Bulletin of the Ministry of Health and the Public Health Service Laboratory* **17**: 270–3.

Anon. (1958b) Pencillin fallout. *Time Magazine* (17 November): 50–1.

Anon. (1960) Editorial: a new penicillin. *British Medical Journal* **ii**: 720–1.

Anon. (1961) Editorial: 'Celbenin'-resistant staphylococci. *British Medical Journal* **i**: 113–14.

Anon. (1979) Obituary: Lawrence Paul Garrod. *Lancet* **ii**: 647.

Anon. (2006) Obituary: E S Anderson. *The Times* (27 March).

Avorn J L, Barrett J F, Davey P G, McEwen S A, O'Brien T F, Levy S B, Alliance for the Prudent Use of Antibiotics. (2001) *Antibiotic Resistance: Synthesis of recommendations by expert policy groups.* Geneva: WHO. Freely available at: http://whqlibdoc.who.int/hq/2001/WHO_CDS_CSR_DRS_2001.10.pdf (visited 30 January 2008).

Ayliffe G A J. (1973) Use of antibiotics and resistance. In Geddes A M, Williams J D. (eds) *Current Antibiotic Therapy.* Edinburgh: Churchill Livingstone; 53–63.

Ayliffe G A J. (2007) Obituary: Edward Lowbury. *Independent* (14 August).

Ayliffe G A J, English M P. (2003) *Infection: From miasmas to MRSA.* Cambridge: Cambridge University Press.

Ayliffe G A J, Lilly H A, Lowbury E J L. (1979a) Decline of the hospital staphylococcus: incidence of multiresistant *Staphylococcus aureus* in three Birmingham hospitals. *Lancet* **i**: 538–41.

Ayliffe G A J, Babb J R, Taylor L, Wise R. (1979b) A unit for source and protective isolation in a general hospital. *British Medical Journal* **ii**: 461–5.

Ayliffe G A J, Fraise A P, Geddes A M, Mitchell K. (eds) (2000) *Control of Hospital Infection: A practical handbook*, 4th edn. London: Arnold.

Ayliffe G A J, Green W, Livingston R, Lowbury E J L. (1977) Antibiotic-resistant *Staphylococcus aureus* in dermatology and burn wards. *Journal of Clinical Pathology* **30**: 40–4.

Ballio A, Chain E B, Dentice Di Accadia F, Rolinson G N, Batchelor F R. (1959) Penicillin derivatives of p-aminobenzylpenicillin. *Nature* **183**: 180–1.

Barber M. (1947) Coagulase-positive staphylococci resistant to penicillin. *Journal of Pathology and Bacteriology* **59**: 373–84.

Barber M. (1948) Letter: sensitization of penicillin-resistant staphylococci. *Lancet* **i**: 730.

Barber M. (1961) Methicillin-resistant staphylococci. *Journal of Clinical Pathology* **14**: 385–93.

Barber M. (1964a) Naturally occurring methicillin-resistant staphylococci. *Journal of General Microbiology* **35**: 183–90.

Barber M. (1964b) Methicillin-resistant staphylococci and hospital infection. *Postgraduate Medical Journal* **40** (Suppl.): 178–81.

Barber M, Garrod L P. (1963) *Antibiotic and Chemotherapy*. Edinburgh: E & S Livingstone Ltd.

Barber M, Wildy P. (1958) A study of the antigenic specificity of staphylococcal coagulase in relation to bacteriophage group. *Journal of General Microbiology* **18**: 92–106.

Barber M, Dutton A A, Beard M A, Elmes P C, Williams R. (1960) Reversal of antibiotic resistance in hospital staphylococcal infection. *British Medical Journal* **i**: 11–17.

Batchelor F R, Doyle F P, Nayler J H C, Rolinson G N. (1959) Syntheses of penicillin: 6-aminopenicillanic acid in penicillin fermentations. *Nature* **183**: 257–8.

Bates J. (1997) Epidemiology of vancomycin-resistant enterococci in the community and the relevance of farm animals to human infection. *Journal of Hospital Infection* **37**: 89–101.

Beckman P S, Eickhoff T C. (eds) (1971) *Proceedings of the International Conference on Nosocomial Infections*. Chicago, IL: American Hospital Association.

van Belkum A, Tassios P T, Dijkshoorn L, Haeggman S, Cookson B, Fry N K, Fussing V, Green J, Feil E, Gerner-Smidt P, Brisse S, Struelens M.(2007) Guidelines for the validation and application of typing methods for use in bacterial epidemiology by the European Society of Clinical Microbiology and Infectious Diseases (ESCMID) Study Group on Epidemiological Markers (ESGEM). *Clinical Microbiology and Infection* **13** (Suppl. 3): 1–46.

Bennison W H, Schwabacher H. (1948) Letter: sensitization of penicillin-resistant bacteria. *Lancet* **i**: 885.

Blair J E, Williams R E O. (1961) 'Phage typing of staphylococci. *Bulletin of the World Health Organization* **24**: 771–84.

Borowski J. (1988a) Overview of current staphylococcal problems in Poland. *Journal of Hospital Infection* **11** (Suppl. A): 116–22.

Borowski J. (1988b) Surveillance of MRSA in Poland. *British Journal of Clinical Practice* (Suppl.) **57**: 72–3.

Borowski J, Kamieńska K, Rutecka I. (1964) Letter: Methicillin-resistant staphylococci. *British Medical Journal* **i**: 983.

Borowski J, Jakubicz P, Jakoniuk P, Ziobro J. (1967) [Characteristics of *Staphylococcus aureus* strains isolated from the hospitals of Bialystok province] Polish. *Roczniki Akademii Medycznej im. Juliana Marchlewskiego w Białymstoku* **13**: 189–98.

Brachman P. (ed.) *Proceedings of the International Conference on Nosocomial Infections*, 1970, Chicago, IL: American Hospital Association.

British Society for Antimicrobial Chemotherapy, Working Party. (1994) Hospital antibiotic control measures in the UK: Working Party Report. *Journal of Antimicrobial Chemotherapy* **34:** 21–42.

British Society for Antimicrobial Chemotherapy and the Hospital Infection Society, Combined Working Party. (1995) Guidelines on the control of methicillin-resistant Staphylococcus aureus in the community. Report of a combined Working Party. *Journal of Hospital Infection* **31:** 1–12.

British Society for Antimicrobial Chemotherapy, the Hospital Infection Society and the Infection Control Nurses Association, Combined Working Party. (1998) Revised guidelines for the control of methicillin-resistant *Staphylococcus aureus* infection in hospitals: report of a combined working party. *Journal of Hospital Infection* **39:** 253–90.

British Society of Antimicrobial Chemotherapy, the Hospital Infection Society and the Infection Control Nurses Association, Joint Working Party. (2006) Guidelines for the control and prevention of meticillin-resistant *Staphylococcus aureus* (MRSA) in healthcare facilities. *Journal of Hospital Infection* **63** (Suppl. 1): S1–44. Also cited as Coia *et al.* (2006).

Brown D F J, Reynolds P E. (1980) Intrinsic resistance to β-lactam antibiotics in *Staphylococcus aureus*. *FEBS Letters* **122**: 275–8.

Brown D F, Edwards D I, Hawkey P M, Morrison D, Ridgway G L, Towner K J, Wren M W; Joint Working Party of the British Society for Antimicrobial Chemotherapy; Hospital Infection Society; Infection Control Nurses Association. (2005) Guidelines for the laboratory diagnosis and susceptibility testing of methicillin-resistant *Staphylococcus aureus* (MRSA). *Journal of Antimicrobial Chemotherapy* **56:** 1000–18.

Brown K. (2004) *Penicillin Man. Alexander Fleming and the antibiotic revolution.* Stroud: Sutton Publishing.

Bud R. (2007) *Penicillin: Triumph and tragedy.* Oxford: Oxford University Press.

Budd W. (1984) *On the Causes of Fever (1839): On the Causes and Mode of Propagation of the Common Continued Fevers of Great Britain and Ireland,* Smith D C. (ed.). Baltimore, MD: Johns Hopkins University Press.

Bulloch W. (1929) Alexander Ogston. *Aberdeen University Review* **16:** 97–102.

Callaghan R P, Cohen S J, Stewart G T. (1961) Septicaemia due to colonization of Spitz-Holter valves by staphylococci. Five cases treated with methicillin. *British Medical Journal* **i:** 860–3.

Casewell M W. (1986). Epidemiology and control of the 'modern' methicillin-resistant *Staphylococcus aureus. Journal of Hospital Infection* **7** (Suppl.): 1–11.

Casewell M W, Hill R L R. (1986) The carrier state: methicillin-resistant *Staphylococcus aureus. Journal of Antimicrobial Chemotherapy* **18** (Suppl. A): 1–12.

Casewell M W, Hill R L R. (1989) Mupirocin for eradication of nasal carriage of staphylococci. *Lancet* **i:** 154.

Casewell M W, Hill R L R. (1991) Minimal dose requircments for nasal mupirocin and its role in the control of epidemic MRSA. *Journal of Hospital Infection* **19** (Suppl. B): 35–40.

Casewell M W, Hill R L R, Duckworth G J. (1984) The effect of mupirocin on the nasal carriage of *Staphylococcus aureus.* In Wilkinson D S, Price J D. (eds) *Mupirocin – A novel antibiotic for the treatment of skin infection.* Royal Society of Medicine Congress and Symposium series no. 80. London: RSM, 149–54.

Centers for Disease Control (CDC). (1977) *National Nosocomial Infections Study Report,* Annual Summary 1975. Atlanta, GA: CDC, 13.

Chabbert Y A, Baudens J G. (1962) [Staphylococcus strains naturally resistant to methicillin and 5-methyl-3-phenyl-4-iso-oxazolyl-penicillin] French. *Annales de l'Institut Pasteur* **103**: 222–30.

Chabbert Y A, Baudens J G, Gerbaud G R. (1964) Variations sous l'influence de l'acriflavine et transduction de la resistance a la kanamcine et au chloramphenicol chez les staphylocoques] French. *Annales de l'Institute Pasteur* **107**: 678–90.

Chain E B, Mellows G. (1977) Pseudomonic acid. Part 1. The structure of pseudomonic acid A, a novel antibiotic produced by *Pseudomonas fluorescens. Journal of the Chemical Society, Perkin Transactions 1* (3): 294–309.

Clark A M, Colebrook L, Gibson T, Thomson M L, Foster A. (1943) Penicillin and propamidine in burns: elimination of haemolytic streptococci and staphylococci. *Lancet* **241**: 605–9.

Coia J E, Duckworth G J, Edwards D I, Farrington M, Fry C, Humphreys H, Mallaghan C, Tucker D R. (2006) Guidelines for the control and prevention of meticillin-resistant *Staphylococcus aureus* (MRSA) in healthcare facilities. Joint Working Party of the British Society of Antimicrobial Chemotherapy; Hospital Infection Society; Infection Control Nurses Association. *Journal of Hospital Infection* **63** (Suppl.) 1: S1–44.

Cooke E M, Marples R R. (1987) Letter: Methicillin-resistant *Staphylococcus aureus. British Medical Journal* **294**: 371.

Cookson B, Phillips I. (1990) Methicillin-resistant staphylococci. *Society for Applied Bacteriology Symposium Series* **19**: 55S–70S.

Cooper B S, Stone S P, Kibbler C C, Cookson B D, Roberts J A, Medley G F, Duckworth G J, Lai R, Ebrahim S. (2003) Systematic review of isolation policies in the hospital management of methicillin-resistant *Staphylococcus aureus*: a review of the literature with epidemiological and economic modelling. *Health Technology Assessment* **7**: 1–194. Freely available at: http://www.hta.ac.uk/execsumm/summ739.htm (visited 5 December 2007).

Cosgrove S E, Sakoulas G, Perencevich E N, Schwaber M J, Karchmer A W, Carmeli Y. (2003) Comparison of mortality associated with methicillin-resistant and methicillin-susceptible *Staphylococcus aureus* bacteremia: a meta-analysis. *Clinical Infectious Diseases* **36**: 53–9.

Cox R A, Conquest C, Mallaghan C, Marples R R. (1995) A major outbreak of methicillin-resistant *Staphylococcus aureus* caused by a new 'phage-type (EMRSA-16). *Journal of Hospital Infection* **29**: 87–106.

Cunningham J B, Kernohan W G, Rush T. (2006) Bed occupancy, turnover intervals and MRSA rates in English hospitals. *British Journal of Nursing* **15**: 656–60.

Daly J S, Eliopolous G M, Willey S, Moellering R C. (1988) Mechanism of action and *in vitro* and *in vivo* activities of S-6123, a new oxazolidinone compound. *Antimicrobial Agents and Chemotherapy* **32**: 1341–6.

Dancer S J. (2008) Importance of the environment in meticillin-resistant *Staphylococcus aureus* acquisition: the case for hospital cleaning. *Lancet Infectious Diseases* **8**: 101–13.

Datta N. (1962) Transmissible drug resistance in an epidemic strain of *Salmonella typhimurium*. *Journal of Hygiene* **60**: 301–10.

Datta N. (1969) Penicillin in poultry feed. *British Medical Journal* **iv**: 741.

Davies R R, Noble W C. (1962) Dispersal of bacteria on desquamated skin. *Lancet* **ii**: 1295–7.

Department of Health (DoH). (2000) All hospitals to monitor hospital-acquired infection. *Department of Health Press Release* 2000/0584 [Monday, 16 October 2000]. London: Department of Health. Freely available at: www.dh.gov.uk/en/Publicationsandstatistics/Pressreleases/DH_4007286 (visited 5 December 2007).

DoH. (2001) Surveillance of healthcare-associated infections. *CMO's Update 30* [21 May 2001]. London: Department of Health. Modified copy (9 February 2007) freely available at: www.dh.gov.uk/en/Publicationsandstatistics/Lettersandcirculars/CMOupdate/DH_4003623 (visited 5 December 2007).

DoH. (2003) *Winning ways: working together to reduce healthcare associated infection in England: Report of the Chief Medical Officer.* London: Department of Health. Freely available at www.dh.gov.uk/en/Publicationsandstatistics/ Publications/PublicationsPolicyAndGuidance/DH_4064682 (visited 6 March 2008).

DoH. (2004) Bloodborne MRSA infection rates to be halved by [March] 2008 – Reid. *Department of Health Press Release* 2000 [Thursday, 5 November 2004]. London: Department of Health. Freely available at: www.dh.gov.uk/en/Publicationsandstatistics/Pressreleases/DH_4093533 (visited 5 December 2007).

DoH. (2008) *Clean, Safe Care: Reducing infections and saving lives.* London: Department of Health. Freely available at www.dh.gov.uk/ en/Publicationsandstatistics/Publications/PublicationsPolicyAndGuidance/ DH_081650 (visited 7 February 2008).

Douthwaite A H, Trafford J A P. (1960) A new synthetic penicillin. *British Medical Journal* **ii**: 687–90.

Doyle P. (1993) John Nayler, 1927–93. *Chemistry in Britain* **29**: 531.

Duckworth G J, Charlett A. (2005) Editorial: improving surveillance of MRSA bacteraemia. *British Medical Journal* **331**: 976–7.

Duckworth G J, Lothian J L, Williams J D. (1988) Methicillin-resistant *Staphylococcus aureus*: report of an outbreak in a London teaching hospital. *Journal of Hospital Infection* **11**: 1–15.

Dunstan E J, Main A N, Rowe J. (1995) In hot pursuit of MRSA. *Lancet* **346**: 1639.

Dyke K G. (1969) Penicillinase production and intrinsic resistance to penicillins in methicillin-resistant cultures of *Staphylococcus aureus*. *Journal of Medical Microbiology* **2**: 261–78.

Dyke K G H, Parker M T, Richmond M H. (1970) Penicillinase production and metal-ion resistance in *Staphylococcus aureus* cultures isolated from hospital patients. *Journal of Medical Microbiology* **3**: 125–36.

Elek S D. (1948a) *Analysis of complex flocculating systems by means of diffusion gradients,* PhD thesis, University of London.

Elek S D. (1948b) The recognition of toxicogenic bacterial strains *in vitro*. *British Medical Journal* i: 493–6.

Elek S D, Conen P E. (1957) The virulence of *Staphylococcus pyogenes* for man: a study of the problems of wound infection. *British Journal of Experimental Pathology* **38:** 573–86.

Elek S D, Fleming P C. (1960) A new technique for the control of hospital cross-infection. Experience with BRL1241 in a maternity unit. *Lancet* **ii:** 569–72.

Engemann J J, Carmeli Y, Cosgrove S E, Fowler V G, Bronstein M Z, Trivette S L, Briggs J P, Sexton D J, Kaye K S. (2003) Adverse clinical and economic outcomes attributable to methicillin resistance among patients with *Staphylococcus aureus* surgical site infection. *Clinical Infectious Diseases* **36:** 592–8.

Enright M C, Robinson D A, Randle G, Feil E J, Grundmann H, Spratt B G. (2002) The evolutionary history of methicillin-resistant *Staphylococcus aureus* (MRSA). *Proceedings of the National Academy of Sciences of the USA* **99:** 7687–92.

Eriksen K R. (1964) Methicillin-resistance in *Staphylococcus aureus* apparently developed during treatment with methicillin. *Acta Pathologica et Microbiologica Scandinavica* **61:** 154–5.

European Society of Clinical Microbiology and Infectious Diseases (ESCMID), Study Group on Epidemiological Markers (ESGEM). (2007) Guidelines for the validation and application of typing methods for use in bacterial epidemiology. *Clinical Microbiology and Infection* **13** (Suppl. 3): 1–46. Also cited as van Belkum *et al.* (2007).

Eykyn S, Jenkins C, King A, Phillips I. (1973) Antibacterial activity of cefamandole, a new cephalosporin antibiotic, compared with that of cephaloridine, cephalothin, and cephalexin. *Antimicrobial Agents and Chemotherapy* **3:** 657–61.

Faber V, Jessen O, Rosendal K, Eriksen K R. (1960) Staphylococcal bacteraemia: clinical and bacteriological observations in 201 cases. *British Medical Journal* **ii:** 1832–6.

Fairbrother R W, Williams B L. (1956) Two new antibiotics; antibacterial activity of novobiocin and vancomycin. *Lancet* **268:** 1177–9.

Filip S V, Cavelier F. (2004) A contribution to the nomenclature of depsipeptides. *Journal of Peptide Science* **10**: 115–18.

Fisk R T. (1942) Studies on staphylococci. Occurrence of bacteriophage carriers amongst strains of *Staphylococcus aureus*. Identification of *Staphylococcus aureus* strains by means of bacteriophage. *Journal of Infectious Diseases* **71**: 153–60.

Ford C W, Hamel J C, Wilson D M, Moerman J K, Stapert D, Yancey R J Jr, Hutchinson D K, Barbachyn M R, Brickner S J. (1996) *In vivo* activities of U-100592 and U-100766, novel oxazolidinone antimicrobial agents, against experimental bacterial infections. *Antimicrobial Agents and Chemotherapy* **40**: 1508–13.

French G L, Phillips I. (1997) Resistance. In O'Grady F, Lambert H P, Finch R G, Greenwood D. (eds) *Antibiotics and Chemotherapy*, 7th edn. New York, NY; Edinburgh: Churchill Livingstone, 23–43.

French G L, Otter J A, Shannon K P, Adams N M, Watling D, Parks M J. (2004) Tackling contamination of the hospital environment by methicillin-resistant *Staphylococcus aureus* (MRSA): a comparison between conventional terminal cleaning and hydrogen peroxide vapour decontamination. *Journal of Hospital Infection* **57**: 31–7.

Fuller A T, Mellows G, Woolford M, Banks G T, Barrow K D, Chain E B. (1971) Pseudomonic acid: an antibiotic produced by *Pseudomonas fluorescens*. *Nature* **234:** 416–7.

Garrod L P. (1966) Mary Barber, 1911–65. *Journal of Pathology and Bacteriology* **92**: 603–10.

Gedney J, Lacey R W. (1982) Properties of methicillin-resistant staphylococci now endemic in Australia. *Medical Journal of Australia* **1**: 448–50.

Gemmell C G, Edwards D I, Fraise A P, Gould F K, Ridgway G L, Warren R E; Joint Working Party of the British Society for Antimicrobial Chemotherapy, Hospital Infection Society and Infection Control Nurses Association. (2006) Guidelines for the prophylaxis and treatment of methicillin-resistant *Staphylococcus aureus* (MRSA) infections in the UK. *Journal of Antimicrobial Chemotherapy* **57**: 589–608.

General Medical Council, Education Committee. (1993) *Tomorrow's Doctors: Recommendations on undergraduate medical education.* Issued by the Education Committee of the General Medical Council in pursuance of section 5 of the Medical Act, 1983. London: General Medical Council. Rev. edn, 2002, available online at: www.gmc-uk.org/education/undergraduate/GMC_tomorrows_doctors.pdf (visited 28 January 2008).

Gillespie W A, Alder V G. (1952) Production of opacity in egg yolk media by coagulase-positive staphylococcus. *Journal of Pathology and Bacteriology* **64**: 187–200.

Gillespie W A, Alder V G, Ayliffe G A J, Powell D E B, Wypkema W. (1961) Control of staphylococcal cross-infection in surgical wards. *Lancet* **i**: 1299–1303.

Gillespie W J, Walenkamp G. (2001) Antibiotic prophylaxis for surgery for proximal femoral and other closed long bone fractures. *Cochrane Database of Systematic Reviews* **1**: CD000244. DOI: 10.1002/14651858.CD000244.

Gould I M. (2005) Control of methicillin-resistant *Staphylococcus aureus* in the UK. *European Journal of Clinical Microbiology and Infectious Diseases* **24**: 789–93. Comment in: *European Journal of Clinical Microbiology and Infectious Diseases* (2005) **24**: 777–9.

Gould I M. (ed.) (2007) *MRSA in Practice.* London: Royal Society of Medicine.

Gould I M, van der Meer J W M. (eds) (2005) *Antibiotic Policies: Theory and practice.* New York, NY; London: Kluwer Academic/Plenum.

Gould J C. (1958) Environmental penicillin and penicillin-resistant *Staphylococcus aureus. Lancet* **i**: 489–93.

Greenwood D, O'Grady F. (1969) Antibiotic-induced surface changes in microorganisms demonstrated by scanning electron microscopy. *Science* **163**: 1076–8.

Gregory W A, Brittelli D R, Wang C L, Kezar H S 3rd, Carlson R K, Park C H, Corless P F, Miller S J, Rajagopalan P, Wuonola M A, McRipley R J, Eberly V S, See A M, Forbes M. (1990) Antibacterials: synthesis and structure-activity studies of 3-aryl-2-oxooxazolidines. 2. The A group. *Journal of Medicinal Chemistry* **33**: 2569–78.

Griffith R S. (1981) Introduction to vancomycin. *Reviews of Infectious Diseases* **3** (Suppl.): 200–4.

Griffiths C, Lamagni T L, Crowcroft N S, Duckworth G J, Rooney C. (2004) Trends in MRSA in England and Wales: analysis of morbidity and mortality data for 1993–2002. *Health Statistics Quarterly* **21**: 15–22.

Hakenbeck R, Tarpay M, Tomasz A. (1980) Multiple changes of penicillin-binding proteins in penicillin-resistant clinical isolates of *Streptococcus pneumoniae*. *Antimicrobial Agents and Chemotherapy* **17**: 364–71.

Hamilton-Miller J M T, Ramsay J. (1967) Stability of cephaloridine and cephalothin to staphylococcal penicillinase. *Journal of General Microbiology* **49**: 491–501.

Hanifah Y A, Hiramatsu K. (1994) Pulsed-field gel electrophoresis of chromosomal DNA of methicillin-resistant *Staphylococcus aureus* associated with nosocomial infections. *Malaysian Journal of Pathology* **16**: 151–6.

Hardman J G, Limbird L E. (eds) (2001) *Goodman & Gilman's The Pharmacological Basis of Therapeutics,* 10th edn. New York, NY: McGraw-Hill.

Hare R. (1983) The scientific activities of Alexander Fleming, other than the discovery of penicillin. *Medical History* **27**: 347–72.

Hare R, Thomas C G. (1956) The transmission of *Staphylococcus aureus*. *British Medical Journal* **ii**: 840–4.

Hartman B J, Tomasz A. (1984) Low-affinity penicillin-binding protein associated with β-lactam resistance in *Staphylococcus aureus*. *Journal of Bacteriology* **158**: 513–16.

Hartstein A I, Denny M A, Morthland V H, LeMonte A M, Pfaller M A. (1995) Control of methicillin-resistant *Staphylococcus aureus* in a hospital and an intensive care unit. *Infection Control and Hospital Epidemiology* **16**: 405–11.

Health Protection Agency (HPA). (2007) *Third Report of the Mandatory Surveillance of Surgical Site Infection in Orthopaedic Surgery, April 2004 to March 2007*. London: Health Protection Agency. Freely available at: www.hpa.org.uk/infections/topics_az/surgical_site_infection/documents/SSI3rdMandatory01-11-07.pdf (visited 5 December 2007).

HPA, Evaluations and Standards Laboratory, Standards Unit. (2006) *Susceptibility Testing.* National Standard Method, BSOP 45, Issue 2 (Issue date: 30.10.06). Freely available from www.hpa-standardmethods.org.uk/documents/bsop/pdf/bsop45.pdf (visited 14 February 2008).

HPA, Evaluations and Standards Laboratory, Standards Unit. (2007a) *Staining Procedures.* National Standard Method Issue 1. Freely available at www.hpa-standardmethods.org.uk/pdf_sops.asp (visited 12 February 2008).

HPA, Evaluations and Standards Laboratory, Standards Unit. (2007b) *Investigation of specimens for Screening for MRSA*, Standard Operating Procedure, BSOP 29, Issue 5. Freely available from www.hpa-standardmethods.org.uk/pdf_sops.asp (visited 7 January 2008).

Hill A B. (1934) *The Inheritance of Resistance to Bacterial Infection in Animal Species: A review of the published experimental data.* Medical Research Council (Great Britain) Special report series no. 196. London: HMSO.

Hill R L R, Duckworth G J, Casewell M W. (1988) Elimination of nasal carriage of methicillin-resistant *Staphylococcus aureus* with mupirocin during a hospital outbreak. *Journal of Antimicrobial Chemotherapy* **22**: 377–84.

Holt R J, Stewart G T. (1964) Penicillin amidase from coliforms: its extraction and some characteristics. *Nature* **201**: 824.

Hookey J V, Richardson J F, Cookson B D. (1998) Molecular typing of *Staphylococcus aureus* based on PCR restriction fragment length polymorphism and DNA sequence analysis of the coagulase gene. *Journal of Clinical Microbiology* **36**: 1083–9.

Hospital Infection Society and British Society for Antimicrobial Chemotherapy, Combined Working Party. (1986) Guidelines for the control of epidemic methicillin-resistant *Staphylococcus aureus.* Report. *Journal of Hospital Infection* 7: 193–201. Revised in 1990 [*Journal of Hospital Infection* **16**: 351–77].

Hospital Infection Society and the British Society for Antimicrobial Chemotherapy, Combined Working Party. (1990) Revised guidelines for the control of epidemic methicillin-resistant Staphylococcus aureus: report prepared by G Duckworth, Secretary to the working party. *Journal of Hospital Infection* **16:** 351–77.

House of Commons, Committee of Public Accounts. (2000) *Forty-second Report: The management and control of hospital-acquired infection in acute NHS trusts in England.* Session 1999/2000 [8 November 2000]. London: Stationery Office. Freely available at: www.publications.parliament.uk/pa/cm199900/cmselect/cmpubacc/306/30603.htm#n1 (visited 5 December 2007).

House of Lords, Select Committee on Science and Technology. (1998) *Resistance to Antibiotics and Other Antimicrobial Agents: Report* of the House of Lords Select Committee on Science and Technology. House of Lords paper, 81-I, 7th report, Session 1997/8. London: The Stationery Office. See www.publications.parliament.uk/pa/ld199798/ldhansrd/vo981116/text/81116-14.htm (visited 14 August 2007).

Hurlimann-Dalel R L, Ryffel C, Kayser FH, Berger-Bachi B. (1992) Survey of the methicillin resistance-associated genes *mecA, mecR1-mecI,* and *femA-femB* in clinical isolates of methicillin-resistant *Staphylococcus aureus. Antimicrobial Agents and Chemotherapy* **36**: 2617–21.

Ingham H R. (2004) Letter: role of the isolation unit in MRSA control. *Journal of Hospital Infection* **58**: 161.

Jevons M P. (1961) Letter: 'Celbenin'-resistant staphylococci. *British Medical Journal* i: 124–5.

Kagan B M, Martin E R, Stewart G T. (1964) *L form* induction of naturally occurring methicillin-resistant strains of *Staphylococcus aureus. Nature* **203**: 1031–3.

Keith S T. (1981) Inventions, patents and commercial development from governmentally financed research in Great Britain: the origins of the National Research Development Corporation. *Minerva* **19**: 92–122.

Kinmonth J B, Hare R, Tracy G D, Thomas C G, Marsh J D, Jantet G H. (1958) Studies of theatre ventilation and surgical wound infection. *British Medical Journal* ii: 407–11.

Kirby W M M. (1944) Extraction of a highly potent penicillin inactivator from penicillin-resistant staphylococci. *Science* **99**: 452–3.

Kirst H A, Thompson D G, Nicas T I. (1998) Historical yearly usage of vancomycin. *Antimicrobial Agents and Chemotherapy* **42**: 1303–4.

Knox R. (1960) A new penicillin (BRL 1241) active against penicillin-resistant staphylococci. *British Medical Journal* ii: 690–3.

Knox R. (1961) Letter: 'Celbenin'-resistant staphylococci. *British Medical Journal* i: 124–5.

Knox R. (1966) Professor Mary Barber. *Nature* **209**: 559.

Knox R, Smith J T. (1963) Stability of methicillin and cloxacillin to staphylococcal penicillinase. *British Medical Journal* ii: 205–7.

Knudsen E T, Rolinson G N. (1960) Absorption and excretion of a new antibiotic (BRL 1241). *British Medical Journal* ii: 700.

Lacey R W. (1987) The primordial MRSA? *Journal of Hospital Infection* **9**: 89–90.

Lamb H M, Figgitt D P, Faulds D. (1999) Quinupristin/dalfopristin: a review of its use in the management of serious Gram-positive infections. *Drugs* **58**: 1061–97.

Lambert H P. (1976) Birthday greeting to LPG[arrod]. *Journal of Antimicrobial Chemotherapy* **2**: 1.

LaPlante K L, Rybak M J. (2004) Daptomycin – a novel antibiotic against Gram-positive pathogens. *Expert Opinion on Pharmacotherapy* **5**: 2321–31.

Lazell H G. (1975) *From Pills to Penicillin: The Beecham story, a personal account.* London: Heinemann.

Lidwell O M. (1959) Apparatus for 'phage typing of *Staphylococcus aureus. Monthly Bulletin of the Ministry of Health and Public Health Laboratory Service* **18**: 49–52.

Lidwell O M, Davies J, Payne R W, Newman P, Williams R E O. (1971) Nasal acquisition of *Staphylococcus aureus* in partly divided wards *Journal of Hygiene* **69:** 113–23.

Lidwell O M, Lowbury E J L, Whyte W, Blowers R, Stanley S J, Lowe D. (1982) Effect of ultraclean air in operating rooms on deep sepsis in the joint after total hip or knee replacement: a randomised study. *British Medical Journal* **285**: 10–14.

Lidwell OM, Lowbury E J L, Whyte W, Blowers R, Stanley SJ, Lowe D. (1984) Infection and sepsis after operations for total hip or knee joint replacement: influence of ultraclean air, prophylactic antibiotics and other factors. *Journal of Hygiene* **93**: 505–29.

Lidwell O M, Polakoff S, Jevons M P, Parker M T, Shooter R A, French V I, Dunkerley D R. (1966). Staphylococcal infection in thoracic surgery: experience in a subdivided ward. *Journal of Hygiene* **64**: 321–7.

Lidwell O M, Polakoff S, Davies J, Hewitt J H, Shooter R A, Walker K A, Gaya H, Taylor G W. (1970) Nasal acquisition of *Staphylococcus aureus* in a subdivided and mechanically ventilated ward: endemic prevalence of a single staphylococcal strain. *Journal of Hygiene* **68**: 417–33.

Lilly H A, Lowbury E J L, Wilkins M D, Cason J S. (1979) Staphylococcal sepsis in a burns unit. *Journal of Hygiene* **83**: 429–35.

Lister J. (1867) On the antiseptic principle in the practice of surgery. *Lancet* **ii**: 353–7, 668.

Loeb M, Main C, Walker-Dilks C, Eady A. (2003) Antimicrobial drugs for treating methicillin-resistant *Staphylococcus aureus* colonization. *Cochrane Database of Systematic Reviews* (**4**):CD003340.

Loveday H P, Pellowe C M, Jones S R, Pratt R J.(2006) A systematic review of the evidence for interventions for the prevention and control of meticillin-resistant *Staphylococcus aureus* (1996–2004): report to the Joint MRSA Working Party (Subgroup A). *Journal of Hospital Infection* **63** (Suppl. 1): S45–70.

Lowbury E J L. (1983) Obituary: Leonard Colebrook (1883–1967). *British Medical Journal* **287**: 1981–3.

Lowbury E J L, Ayliffe G A J. (1974) *Drug Resistance in Antimicrobial Therapy*. Springfield, IL: Thomas.

Lowbury E J L, Lilly H A, Kidson A. (1977) 'Methicillin-resistant' *Staphylococcus aureus*: reassessment by controlled trial in burns unit. *British Medical Journal* **i**: 1054–6.

Lowbury E J L, Ayliffe G A J, Geddes A M, Williams J D. (eds) (1975) *Control of Hospital Infection: A practical handbook*. London: Chapman and Hall for the Working Party on Control of Hospital Infection. Later editions edited by Professor Graham Ayliffe, for example, Ayliffe *et al.* (2000).

Lyell A. (1989) Alexander Ogston, micrococci and Joseph Lister. *Journal of the American Academy of Dermatology* **20**: 302–10.

Maiden M C, Bygraves J A, Feil E, Morelli G, Russell J E, Urwin R, Zhang Q, Zhou J, Zurth K, Caugant D A, Feavers I M, Achtman M, Spratt B G. (1998) Multilocus sequence typing: a portable approach to the identification of clones within populations of pathogenic microorganisms. *Proceedings of the National Academy of Sciences of the USA* **95**: 3140–5.

Maple P A, Hamilton-Miller J M, Brumfitt W. (1991) Differing activities of quinolones against ciprofloxacin-susceptible and ciprofloxacin-resistant, methicillin-resistant *Staphylococcus aureus*. *Antimicrobial Agents and Chemotherapy* **35**: 345–50.

Marples R R. (1981) Taxonomic studies of staphylococci and micrococci. *Zentralblattfuer Bakteriologie* **10** (Suppl.): 9–13.

Marples R R, Cooke E M. (1988) Current problems with methicillin-resistant *Staphylococcus aureus. Journal of Hospital Infection* **11**: 381–92.

Marples R R, Richardson J F. (1982) Evaluation of a micromethod gallery (API *Staph*) for the identification of staphylococci and micrococci. *Journal of Clinical Pathology* **35**: 650–6.

Medical Research Council (MRC), Cross Infection Committee. (1951) *The Control of Cross-infection in Hospitals*. London: HMSO.

Ministry of Health. (1959) *Staphylococcal Infections in Hospital:* Report of the Subcommittee of the Central Health Services Council. London: HMSO.

Mitchison J M. (1991) Michael Meredith Swann, Baron Swann of Coln Denys. *Biographical Memoirs of Fellows of the Royal Society* **37**: 446–60.

National Audit Office. (2000) *Report by the Comptroller and Auditor General: The management and control of hospital acquired infection in acute NHS Trusts in England.* HC 230, Session 1999/2000 [February 2000]. London: The Stationery Office.

National Audit Office. (2004) *Report by the Comptroller and Auditor General. Improving patient care by reducing the risk of hospital acquired infection: a progress report.* HC 876, Session 2003/4 [14 July 2004]. London: The Stationery Office.

Naughton P. (2005) Tories dismiss MRSA figures as 'trickery', *The Times Online*, (7 March).

Newsom S W B. (2004a) MRSA and its predecessor – a historical view. Part three: the rise of MRSA and EMRSA. *British Journal of Infection Control* **5**: 25–8.

Newsom S W B. (2004b) MRSA – past, present, future. *Journal of the Royal Society of Medicine* **97**: 509–10.

Nightingale F. (1863) *Notes on Hospitals*, 3rd edn. London: Longman, Green, Longman, Roberts, and Green.

Nixon H H, O'Donnell B. (1961) *The Essentials of Pædiatric Surgery.* London: W. Heinemann (Medical).

Noble W C, Virani Z, Cree R G. (1992) Co-transfer of vancomycin and other resistance genes from *Enterococcus faecalis* NCTC 12201 to *Staphylococcus aureus. FEMS Microbiology Letters* **72**: 195–8.

Oeding P. (1952) Serological typing of staphylococci. *Acta Pathologica et Microbiologica Scandinavic* **93** (Suppl.): 356–63.

Oeding P, Williams R E O. (1958) The type classification of *Staphylococcus aureus.* A comparison of 'phage typing with serological typing. *Journal of Hygiene* **56**: 445–54.

Ogston A. (1881) Report upon micro-organisms in surgical diseases. *British Medical Journal* **i**: 369–74.

Ogston A. (1984) Classics in infectious diseases. On abscesses. Alexander Ogston (1844–1929). *Reviews of Infectious Diseases* **6**: 122–8.

Ouchterlony O. (1948) *In vitro* method for testing toxin-producing capacity of diphtheria bacteria. *Acta Pathologica et Microbiologica Scandinavica* **25**: 186–91.

Pal S C, Ray B G. (1964) Methicillin-resistant staphylococci. *Journal of the Indian Medical Association* **42**: 512–17.

Panton P N, Valentine F C O. (1932) Staphylococcal toxin. *Lancet* **i**: 506–8.

Parker M T. (1962) 'Phage-typing and the epidemiology of *Staphylococcus aureus* infections. *Journal of Applied Bacteriology* **25**: 389–402.

Parker M T. (1971) Methicillin-resistant staphylococci. In Brachman P S, Eickhoff T C. (eds) *Proceedings of the International Conference on Nosocomial Infections 1970*, Chicago, IL: American Hospital Association, 112–16.

Parker M T, Jevons M P. (1963) Hospital strains of staphylococci. In Williams R E O, Shooter R A (eds) *Infection in Hospitals, Epidemiology and Control.* Oxford: Blackwell Scientific Publications, 5–66.

Parker M T, Ashehov E H, Hewitt J H, Nakhla L S, Brock B M. (1974) Endemic staphylococcal infection in hospitals. *Annals of the New York Academy of Sciences* **236**: 466–84.

Peard M C, Fleck D G, Garrod L P, Waterworth P M. (1970) Combined rifampicin and erythromycin for bacterial endocarditis. *British Medical Journal* **iv**: 410–11.

Phillips I. (1991) Epidemic potential and pathogenicity in outbreaks of infection with EMRSA and EMREC. *Journal of Hospital Infection* **18** (Suppl. A): 197–201.

Phillips I. (2007) MRSA: a historical perspective. In Gould I. (ed.) *MRSA in Practice.* London: Royal Society of Medicine: 1–12.

Pirson M, Dramaix M, Struelens M, Riley T V, Leclercq P. (2005) Costs associated with hospital-acquired bacteraemia in a Belgian hospital. *Journal of Hospital Infection* **59**: 33–40.

Pitt T. (2006) Obituary: Dr Tom Parker. *Laboratory of HealthCare Associated Infection Newsletter* **1**: 3.

Pollock M R. (1965) Mary Barber. *Journal of Clinical Pathology* **18**: 697–8.

Public Health Laboratory Service (PHLS). (1960) Incidence of surgical wound infection in England and Wales. *Lancet* **ii**: 659.

Radford T. (1998) Obituary: Anthony Tucker. *Guardian* (16 September): 20.

Rahman M, Sanderson P J, Bentley A H, Barrett S P, Karim Q N, Teare E L, Chaudhuri A, Alcock S R, Corcoran G D, Azadian B, Dance D A, Gaunt P N, Cunningham R, Ahmad F J, Garvey R J, Chattopadhyay B, Wiggins R J, Sheppard M, Wright E P, Moulsdale M, Falkiner F. (2000) Control of MRSA. *Journal of Hospital Infection* **44**: 151–3.

Reeves D S, Phillips I, Williams J D, Wise R. (eds) (1978) *Laboratory Methods in Antimicrobial Chemotherapy.* Edinburgh: Churchill Livingstone.

Reynolds L A, Tansey E M. (eds) (2007a) Early development of total hip replacement. *Wellcome Witnesses to Twentieth Century Medicine*, vol. 29. London: The Wellcome Trust Centre for the History of Medicine at UCL. Freely available online at www.ucl.ac.uk/histmed/publications/wellcome-witnesses/index.html or following the links to Publications/Wellcome Witnesses from www.ucl.ac.uk/histmed.

Reynolds L A, Tansey E M. (eds) (2007b) Medical ethics education in Britain, 1963–93. *Wellcome Witnesses to Twentieth Century Medicine,* vol. 31. London: The Wellcome Trust Centre for the History of Medicine at UCL.

Richards H C, Housley J R, Spooner D F. (1963) Quinacillin: a new penicillin with unusual properties. *Nature* **199**: 354–6.

Richardson J F, Marples R R. (1982) Changing resistance to antimicrobial drugs and resistance typing in clinically significant strains of *Staphylococcus epidermidis. Journal of Medical Microbiology* **15**: 475–84.

Ridgway G, Stokes E J. (2005) Pamela M Waterworth, August 1920–July 2004. *British Society for Antimicrobial Chemotherapy Newsletter* **5**: 3.

Ritchie K, Bradbury I, Craig J, Eastgate J, Foster L, Kohli H, Iqbal K, MacPherson K, McCarthy T, McIntosh H, Nic Lochlainn E, Reid M, Taylor J. (2007) *Health Technology Assessment Report 9: The clinical and cost effectiveness of screening for meticillin-resistant* Staphylococcus aureus *(MRSA).* Glasgow; Edinburgh: NHS Quality Improvement Scotland. Freely available at: www.nhshealthquality.org/nhsqis/files/PatientSafety_HTA9_MRSA_Oct07.pdf (visited 17 January 2008).

Roberts R B, de Lencastre A, Eisner W, Severina E P, Shopsin B, Kreiswirth B N, Tomasz A. (1998) Molecular epidemiology of methicillin-resistant *Staphylococcus aureus* in 12 New York hospitals. *Journal of Infectious Diseases* **178**: 164–71.

Roberts R B, Chung M, de Lencastre H, Hargrave J, Tomasz A, Nicolau D P, John J F, Korzeniowski O; Tri-State MRSA Collaborative Study Group. (2000) Distribution of methicillin-resistant *Staphylococcus aureus* clones among health care facilities in Connecticut, New Jersey, and Pennsylvania. *Microbial Drug Resistance* **6**: 245–51.

Robinson D A, Kearns A M, Holmes A, Morrison D, Grundmann H, Edwards G, O'Brien F G, Tenover F C, McDougal L K, Monk A B, Enright M C. (2005) Re-emergence of early pandemic *Staphylococcus aureus* as a community-acquired meticillin-resistant clone. *Lancet* **365**: 1256–8.

Rolinson G N. (1961) 'Celbenin'-resistant staphylococci. *British Medical Journal* **i**: 125–6.

Rolinson G N. (1998) Historical perspective: forty years of β-lactam research. *Journal of Antimicrobial Chemotherapy* **41**: 589–603.

Rolinson G N, Geddes A M. (2007) The 50th anniversary of the discovery of 6-aminopenicillanic acid (6-APA). *International Journal of Antimicrobial Agents* **29**: 3-8. Erratum in: *International Journal of Antimicrobial Agents* (2007) **29**: 613.

Rolinson G N, Stevens S, Batchelor F R, Wood J C, Chain E B. (1960) Bacteriological studies on a new penicillin – BRL 1241. *Lancet* **ii**: 564–7.

Rosenbach F J. (1884) *Mikro-organismen bei den Wund-infektions-krankheiten des Menschen.* Wiesbaden: J F Bergmann.

Rosendal K. (1971) Current national patterns – Denmark. In Beckman P S, Eickhoff T C. (eds) *Proceedings of the International Conference on Nosocomial Infections.* Chicago, IL: American Hospital Association, 11–16.

Rosendal K, Jessen O, Bentzon M W, Bülow P. (1977) Antibiotic policy and spread of *Staphylococcus aureus* strains in Danish hospitals, 1969–74. *Acta Pathologica Microbiologica Scandinavica, Section B* **85**: 143–52.

Rountree P M, Barbour R G. (1950) *Staphylococcus pyogenes* in new-born babies in a maternity hospital. *Medical Journal of Australia* **1**: 525–8.

Rountree P M, Barbour R G. (1951) Incidence of penicillin-resistant and streptomycin-resistant staphylococci in a hospital. *Lancet* **i**: 435–6.

Rountree P M, Freeman B M. (1955) Infections caused by a particular 'phage type of *Staphylococcus aureus. Medical Journal of Australia* **42**: 157–61.

Salgado C D, Farr B M, Calfee D P. (2003) Community-acquired methicillin-resistant *Staphylococcus aureus*: a meta-analysis of prevalence and risk factors. *Clinical Infectious Diseases* **36**: 131–9.

Sanderson D, Gibb F. (2008) Leslie Ash gets £5m payout from hospital where she caught MRSA. *The Times* (17 January).

Sato K, Shiratori O, Katagiri K. (1967) The mode of action of quinoxaline antibiotics. Interaction of quinomycin A with deoxyribonucleic acid. *Journal of Antibiotics* **20**: 270–6.

Selkon J B, Stokes E R, Ingham H R. (1980) The role of an isolation unit in the control of hospital infection with methicillin-resistant staphylococci. *Journal of Hospital Infection* **1**: 41–6.

Shanson D C. (1981) Antibiotic-resistant *Staphylococcus aureus. Journal of Hospital Infection* **2**: 11–36.

Shanson D C. (1982) *Microbiology in Clinical Practice.* Bristol: J Wright.

Shanson D C, McSwiggan D A. (1980) Operating theatre-acquired infection with a gentamicin-resistant, methicillin-resistant strain of *Staphylococcus aureus:* outbreaks in two hospitals attributable to one surgeon. *Journal of Hospital Infection* **1**: 171–2.

Shanson D C, Johnstone D, Midgley J. (1985) Control of a hospital outbreak of methicillin-resistant *Staphylococcus aureus* infections: value of an isolation unit. *Journal of Hospital Infection* **6**: 285–92.

Shanson D C, Kensit J G, Duke R. (1976) Outbreak of hospital infection with a strain of *Staphylococcus aureus* resistant to gentamicin and methicillin. *Lancet* **ii**: 1347–8.

Sheldrick G M, Heine A, Schmidt-Base K, Pohl E, Jones P G, Paulus E, Waring M J. (1995) Structures of quinoxaline antibiotics. *Acta Crystallographica. Section B: Structural Crystallography and Crystal Chemistry* **51**: 987–99.

Shooter R A, Smith M A, Griffiths J D, Brown M E, Williams R E O, Rippon J E, Jevons M P. (1958) Spread of staphylococci in a surgical ward. *British Medical Journal* **i**: 607–13.

Shorr A F. (2007) Epidemiology of staphylococcal resistance. *Clinical Infectious Diseases* **45** (Suppl. 3): S171–6.

Sievert D M, Boulton M L, Stoltman G, Johnson D, Stobierski M G, Downes F P, Somsel P A, Rudrik J T, Brown W, Hafeez W, Lundstrom T, Flanagan E, Johnson R, Mitchell J, Chang S. (2002) *Staphylococcus aureus*-resistant to vancomycin, US, 2002. *Morbidity and Mortality Weekly Report* (5 July 2002) **51**: 565–7.

Simpson E. (1988) Sir Peter Medawar 1915–1987. *Immunology Today* **9**: 4–6.

Slee A M, Wuonola M A, McRipley R J, Zajac I, Zawada M J, Bartholomew P T, Gregory W A, Forbes M. (1987) Oxazolidinones, a new class of synthetic antibacterial agents: *in vitro* and *in vivo* activities of DuP 105 and DuP 721. *Antimicrobial Agents and Chemotherapy* **31**: 1791–7.

Song M D, Wachi M, Doi M, Ishino F, Matsuhashi M. (1987) Evolution of an inducible penicillin-target protein in methicillin-resistant *Staphylococcus aureus* by gene fusion. *FEBS Letters* **221**: 167–71.

Speller D C, Raghunath D, Stephens M, Viant A C, Reeves D S, Wilkinson P J, Broughall J M, Holt H A. (1976) Epidemic infection by a gentamicin-resistant *Staphylococcus aureus* in three hospitals. *Lancet* **i**: 464–6.

Spicer W J. (1984) Three strategies in the control of staphylococci including methicillin-resistant *Staphylococcus aureus*. *Journal of Hospital Infection* **5** (Suppl. A): 45–9.

Spink W W. (1951) Clinical and biologic significance of penicillin-resistant staphylococci, including observations with streptomycin, aureomycin, chloramphenicol, and terramycin. *Journal of Laboratory and Clinical Medicine* **37**: 278–93.

Spink W W. (1962) Pathogenesis and therapy of shock due to infection, experimental and clinical studies. In Bock K D. (ed.) *Shock: Pathogenesis and therapy: an international symposium.* Berlin: Springer; New York, NY: Academic Press.

Spratt B G. (1975) Distinct penicillin-binding proteins involved in the division, elongation, and shape of *Escherichia coli* K12. *Proceedings of the National Academy of Sciences of the USA* **72**: 2999–3003.

Spratt B G, Pardee A B. (1975) Penicillin-binding proteins and cell shape in *E. coli. Nature* **254**: 516–17.

Stern H, Elek S D. (1955) Combined antibiotic therapy for the suppression of resistant variants in urinary infection. *British Medical Journal* **ii**: 1304–6.

Stewart G T. (1946) The effect of penicillin upon Gram-negative bacteria. *Journal of Hygiene* **45**: 282–8.

Stewart G T. (1965) *The Penicillin Group of Drugs*. Amsterdam; New York, NY: Elsevier.

Stewart G T. (1970) Epidemiological approach to assessment of health. *Lancet* **ii**: 115–19.

Stewart G T. (1992) Changing case-definition for AIDS. *Lancet* **340**: 1414.

Stewart G T, Holt R J. (1963) Evolution of natural resistance to the newer penicillins. *British Medical Journal* **i**: 308–11.

Stewart G T, Coles H M T, Nixon H H, Holt R J. (1961) 'Penbritin': an oral penicillin with broad-spectrum activity. *British Medical Journal* **ii**: 200–6.

Stewart G T, Nixon H H, Coles H M, Kesson C W, Lawson D, Thomas R G, Mishra J N, Mitchell M E, Semmens J M, Wade T H. (1960) Report on clinical use of BRL 1241 in children with staphylococcal and streptococcal infections. *British Medical Journal* **ii**: 703–6.

Sutherland R, Rolinson G N. (1964) Characteristics of methicillin-resistant Staphylococci. *Journal of Bacteriology* **87**: 887–99.

Tansey E M, Reynolds L A. (eds) (2000) Post penicillin antibiotics: From acceptance to resistance? *Wellcome Witnesses to Twentieth Century Medicine*, vol. 6. London: The Wellcome Trust. Freely available online at www.ucl. ac.uk/histmed/publications/wellcome-witnesses/index.html or following the links to Publications/Wellcome Witnesses from www.ucl.ac.uk/histmed.

Tarr H A. (1958) Mechanical aids for the 'phage-typing of *Staphylococcus aureus*. *Monthly Bulletin of the Ministry of Health and the Public Health Service Laboratory* **17**: 64–72.

Tucker A. (1978) The brave new world of test tube babies: not all scientists welcome the advent of human embryos conceived in the laboratory. *Guardian* (27 July). Freely available at www.guardian.co.uk/environment/1986/apr/29/energy.russia (visited 12 February 2008).

Tucker A. (1986) Radioactive Russian dust cloud escapes: major nuclear power accident reported at Chernobyl plant in the Soviet Union. *Guardian* (29 April). Freely available at www.guardian.co.uk/medicine/story/0,,1005345,00.html (visited 12 February 2008).

Tucker A. (2006) Obituary: E S Anderson: brilliant bacteriologist who foresaw the public health dangers of genetic resistance to antibiotics. *Guardian* (22 March). Freely available at http://education.guardian.co.uk/higher/news/story/0,,1736615,00.html (visited 12 February 2008).

US Department of Health Education and Welfare. (1958) *On Hospital-acquired Staphylococcal Disease:* Proceedings of the National Conference on Hospital-acquired Staphylococcal Disease held at Atlanta, Georgia, 15–17 September 1958, sponsored by the US Public Health Service, the Communicable Disease Center, the National Academy of Sciences and the National Research Council. Atlanta, GA: Communicable Disease Center.

Wasserman H H. (2006) Chemistry: synthesis with a twist. *Nature* **441**: 699–700.

Waterworth P M. (1978) Quantitative methods for bacterial sensitivity testing. In Reeves D S, Phillips I, Williams J D, Wise R. (eds) *Laboratory Methods in Antimicrobial Chemotherapy.* Edinburgh: Churchill Livingstone, 31–40.

Watts G. (2006) Obituary: Ephraim Saul Anderson. *Lancet* **367**: 1392.

Whitby M, McLaws M L, Berry G. (2001) Risk of death from methicillin-resistant *Staphylococcus aureus* bacteraemia: a meta-analysis. *Medical Journal of Australia* **175**: 264–7. Comment in: *Medical Journal of Australia* (2002) **176** : 188; author reply, 189.

White L F, Dancer S J, Robertson C. (2007) A microbiological evaluation of hospital cleaning methods. *International Journal of Environmental Health Research* **17**: 285–95.

White L F, Dancer S J, Robertson C, MacDonald J (2008) Are hygiene standards useful in assessing infection risk? *American Journal of Infection Control* **36**: forthcoming.

Williams R E O. (1959) Epidemic staphylococci. *Lancet* **i**: 190–5.

Williams R E O. (1963) Healthy carriage of *Staphylococcus aureus*: its prevalence and importance. *Bacteriological Reviews* **27**: 56–71.

Williams R E O, Shooter R A. (eds) (1963) *Infection in Hospitals, Epidemiology and Control.* Oxford: Blackwell Scientific Publications.

Williams R E O, Blowers R, Garrod L P, Shooter R A. (1960) *Hospital Infection: Causes and prevention* [2nd edn, 1966]. London: Lloyd-Luke Medical Books.

Wilson G S, Miles A A. (eds) (1964) *Topley and Wilson's Principles of Bacteriology and Immunity*, 5th edn, 2 vols. London: Edward Arnold Publishers Ltd.

Wise R I, Cranny C, Spink W W. (1956) Epidemiologic studies on antibiotic-resistant strains of *Micrococcus pyogenes*. *American Journal of Medicine* **20**: 176–84.

Witte W. (2000) Selective pressure by antibiotic use in livestock. *International Journal of Antimicrobial Agents* **16** (Suppl.): S19–24.

Woodford N, Johnson A P, Morrison D, Speller D C. (1995) Current perspectives on glycopeptide resistance. *Clinical Microbiology Reviews* **8**: 585–615.

Wyllie D H, Crook D W, Peto T E. (2006) Mortality after *Staphylococcus aureus* bacteraemia in two hospitals in Oxfordshire, 1997–2003: cohort study. *British Medical Journal* **333**: 281, Epub 2006 Jun 23. Erratum in: *BMJ* **333**: 468.

Biographical notes*

Professor Ephraim Saul (Andy) Anderson

CBE FRS (1911–2006) qualified at King's College Medical School, Newcastle (then part of Durham University) and was in general practice until he joined the Royal Medical Corps in 1939 and served in Cairo. After further training in pathology and bacteriology at the Royal Postgraduate Medical School and studies of typhoid 'phage types at the Lister Institute of Preventive Medicine, London, he joined the Enteric Reference Laboratory of the Public Health Laboratory Service, Colindale, in 1947, and was Director from 1954 until his retirement in 1978, as well as the International Reference Laboratory for Enteric 'Phage Typing and the WHO Collaborating Centre on Drug Resistance in Enterobacteria from 1960, both at Colindale. His recognition of the health implications of multiple drug-resistance in bacteria to antibiotics contributed to the formation of the Swann Committee [Agricultural Research Council and Medical Research Council, Joint Committee on Antibiotics in Animal Feeding

(1969)], a risk acknowledged by doctors and governments in the 1990s. Anderson (1968); Datta (1969); Tucker (2006); Anon. (2006); Watts (2006).

Professor Graham Ayliffe

MD FRCPath HonDipHIC (b. 1926) Developed an interest in hospital infection in Bristol with Professor William Gillespie (1956–59). In 1960 he joined Mary Barber at Hammersmith Hospital and was appointed Senior Scientific Officer (MRC) and Hon Consultant with Professor Edward Lowbury at the Hospital Infection Research Laboratory, Birmingham in 1964. He was then appointed Hon Director (1980–94), Director of the WHO Collaborating Centre for Hospital Infection (1985) and Professor of Medical Microbiology, University of Birmingham, 1980–91. He was Chairman of the Combined Working Party on MRSA, 1996–98. See Lowbury *et al.* (eds) (1975).

Professor Mary Barber

MD FIBiol FRCPath (1911–65) was Professor in Clinical Bacteriology at the Postgraduate Medical School, Hammersmith

* Contributors are asked to supply details; other entries are compiled from conventional biographical sources.

Hospital, London, from 1964 until her sudden death in 1965. She had been at the Hammersmith from 1947–48 as lecturer and reader from 1957 to 1964. See Pollock (1965); Garrod (1966); Knox (1966).

Dr Ralph Batchelor

(b. 1931), a biochemist at Beecham Research Laboratories, Brockham Park, Betchworth, Surrey, from 1956, worked his first year in Rome with Professor Sir Ernst Chain. He moved from research to general management in 1970 and was a Director of Beecham Pharmaceuticals from 1978 until his retirement in 1989. He was awarded the Addingham Medal by the City of Leeds in 1966 and Mullard Medal of the Royal Society in 1971. Some of the first crystals and early chromatograms showing the presence of 6-APA are in his original notebook, now on display in the Science Museum. See Batchelor *et al.* (1959). See Figure 1.

Dr Robert Blowers

FRCP FRCPath (1915–2004) qualified at the Middlesex Hospital, London, and trained in blood transfusion before taking command of No 4 Field Transfusion Unit, which served at D-Day. He was the first medical officer to enter concentration camps after their liberation. He trained in pathology

after the war and was appointed to the Staphylococcus Reference Laboratory at the Central Public Health Laboratory, Colindale. In 1951 he became the director of the Middlesbrough public health laboratory and consultant microbiologist to the Teeside Hospital Group and the Newcastle Regional Hospital Board. He went to Uganda as Professor of Medical Microbiology at the Makerere University College, Kampala, Uganda, in 1967, returning to head the Division of Hospital Infection and Microbiology at the Medical Research Council's Clinical Research Centre, Northwick Park Hospital, Harrow, in 1970 until his retirement in 1980.

Dr Robert Bud

FRHS (b. 1952) took his doctorate at the University of Pennsylvania in the History and Sociology of Science. Since 1978 he has been at the Science Museum, where he is currently Principal Curator of Medicine. He was the Joint Winner of the Bunge Prize for the History of Scientific Instruments in 1998. He was responsible for the Science Museum temporary exhibition, 'Penicillin: A Tale of Triumph and Tragedy' (2007/08). See Bud (2007).

Professor Mark Casewell

MD FRCP FRCPath (b. 1940) was Lecturer at St Thomas' Hospital, London (1971–81) and Reader in Medical Microbiology at the Royal London Hospital (1981–84), until appointed Professor of Medical Microbiology at King's College School of Medicine and Dentistry, London, and Head of the Dulwich Public Health Laboratory from 1984 until his retirement in 1997, later Emeritus. He has a 30-year research interest in hospital infection and the epidemiology and control of infection caused by multiply antibiotic-resistant organisms, especially *Klebsiella* species and MRSA. He was the first to identify the activity of mupirocin against MRSA and its role in the control of epidemic MRSA. He was a founder member of the Hospital Infection Society and served as its Chairman (1988–91).

Dr Bilwanath (Bill) Chattopadhyay

DCP FRCPath (b. 1939) was Consultant Medical Microbiologist and Director of the Public Health Laboratory at Whipps Cross University Hospital, London, between 1974 and 2002. He was also Honorary Senior Lecturer in Medical Microbiology, University of London, during that period. Public health microbiology work covered eight boroughs of London in the North East

Thames Region. He was appointed Consultant in Communicable Diseases Control (CCDC) by the borough of Waltham Forest, 1991–93; the North East Thames Regional Adviser in Medical Microbiology for the Royal College of Pathologists during the 1990s. During the same time he was selected as a national inspector in Medical Microbiology for the Clinical Pathology Accreditation (CPA) Board (UK) Limited, which he held for 11 years, and was a Consultant Medical Microbiologist at North Middlesex University Hospital, London, 2002–06. He also worked for several years as a national Expert Witness in medical microbiology and infectious diseases.

Dr Stephanie Dancer

DTM&H MD FRCPath (b. 1959) is a consultant microbiologist at the Southern General Hospital in Glasgow, and the current editor of the *Journal of Hospital Infection.* She trained at St Bartholomew's Hospital in London (1977–83), followed by postgraduate studies in Pathology at Guy's Hospital, where she gained a thesis on the epidemiology of toxin-producing staphylococci. She has worked in various remote areas of the world, including Papua New Guinea, Thailand, Vietnam and the Canadian High Arctic, and published her experiences, albeit with a microbiological slant. She spent six years as the Control of Infection Officer for Argyll before moving to Health Protection Scotland as its inaugural microbiologist (2002–05). At present, she balances clinical and editorial duties with research projects on the control of MRSA, specifically the role of cleaning.

Dr Bernard Dixon

OBE FIBiol (b. 1938) has been European Editor for the American Society for Microbiology since 1997 and a columnist for *Current Biology* since 2000 and for *Lancet Infectious Diseases* since 2001. He was Editor of *New Scientist* from 1969 to 1979. He has run media training and science communication courses for the University of Oxford and Cancer Research UK since 1991; and has been a member of the European Federation of Biotechnology's Task Group on Public Perceptions of Biotechnology. He holds an honorary DSc from Edinburgh University for contributions to public debate on scientific issues and was appointed OBE for services to science journalism. He has received the Charter Award of the Institute of Biology for services to biology and in 2002 shared (with Steven Rose) the Biochemical Society Award 'for

scientific communication in the public domain'.

Dr Peter Doyle

OBE CChem FRSC (1921–2004) was a chemist in industry after his graduation from the University of London in 1944. He joined the Beecham Research Laboratory at Brockham Park, Surrey, in 1952 and was Director of Research of Beecham Pharmaceuticals from 1962 until his retirement in 1983. His other honours include the Gold Medal in Therapeutics of the Worshipful Society of Apothecaries in 1964 awarded jointly with Dr G N Rolinson; and the Mullard Medal of the Royal Society in 1971 jointly with Dr Ralph Batchelor, Dr J H C Nayler and Dr George Rolinson. He was made OBE for services to the pharmaceutical industry in 1977. See Figure 1.

Dr Georgia Duckworth

FRCP FRCPath (b. 1954) studied veterinary medicine at Churchill College in the first year of mixed colleges at Cambridge, later switching to human medicine and embarking on clinical studies in the first intake at Cambridge University School of Clinical Medicine, after a brief sojourn in Oxford to fill in her gaps on human anatomy and specialization in medical microbiology in Professor J D Williams' department at the

London Hospital Medical College, London (1982–97). Her early research interest concentrated on MRSA and she was Honorary Secretary of various national working parties on its control and management. She was one of the first microbiologists to be appointed a Consultant in Communicable Disease Control, while also Senior Lecturer in Medical Microbiology at the London Hospital Medical College in 1989; later as Regional Epidemiologist for North Thames, then London (1997–2000). In 2001, she established the Department of Healthcare-associated Infection and Antimicrobial Resistance in the Communicable Disease Surveillance Centre, then part of the Public Health Laboratory Service, Colindale, London (now the Centre for Infections of the Health Protection Agency, Colindale). She is Deputy Chair of the Steering Group on Healthcare-associated Infection and Expert Adviser on the European Communicable Disease Centre Panel of Scientific Experts. She has served on the Council of the Hospital Infection Society twice.

Professor Brian Duerden
(b. 1948) is the Inspector of Microbiology and Infection Control at the Department of Health (DH) and Professor of Medical Microbiology at Cardiff University. He qualified at Edinburgh University (1972) and was a Lecturer in Bacteriology there and later at Sheffield University, becoming Professor of Medical Microbiology in 1983. He was Consultant Microbiologist to the Sheffield Children's Hospital. In 1991 he became Professor of Medical Microbiology and Director of the Public Health Laboratory in Cardiff. He was Deputy Director and Medical Director of the PHLS in England and Wales from 1995 and Director of the Service from August 2002 until it became part of the Health Protection Agency. He moved to the DH in 2004. His major interests are anaerobic microbiology, healthcare-associated infections and antibiotic resistance and he was Editor-in-Chief of the *Journal of Medical Microbiology* for 20 years (1982–2002).

Professor Michael Emmerson
OBE FRCPath FRCP FMedSci
(b. 1937) received basic training in Microbiology at University College Hospital (UCH), London, under Dr Joan Stokes. He was appointed Professor of Clinical Bacteriology at Queen's University of Belfast

(1984–89), later Professor of Microbiology at University of Leicester (1989–91); and Professor of Microbiology at University of Nottingham (1989–2000), later Emeritus. He co-founded and served as chairman of the Hospital Infection Society and was its President (2002–06).

Sir Alexander Fleming

Kt FRCP FRS (1881–1955) discoverer of penicillin in 1929, was trained at St Mary's Hospital Medical School (1906–08), entering with a scholarship in natural sciences, joining Sir Almroth Wright's team there in 1908, where he was Professor of Bacteriology from 1928 until his retirement in 1948, later Emeritus. He shared the Nobel Prize for Physiology or Medicine in 1945 for the discovery and development of penicillin with Lord Florey OM Kt FRS (1898–1968) and Sir Ernst Chain Kt FRS (1906–1979). Among many honors, he was President of the London Ayrshire Society and Pathological Comparative Medicine Sections, RSM; and the Society for General Microbiology and Rector of Edinburgh University, 1951–54. For a background to his work, see Hare (1983).

Professor Gary French

MD FRCPath DipHIC (b. 1945) qualified in medicine at St Thomas' Hospital, London, and became a Senior Lecturer/Hon. Consultant in Microbiology in 1980. In 1982 he was appointed to the Foundation Chair of Microbiology at the Chinese University of Hong Kong and the Prince of Wales Hospital and in 1990 was made Professor of Microbiology and Honorary Consultant at Guy's and St Thomas' Hospitals (merged into King's College London in 1998). He has been Chairman of the Hospital Infection Society (1999–2001) and Editor of the *Journal of Hospital Infection* (1995–99). He is a member of Government advisory committees on Antimicrobial Resistance and Healthcare Associated Infections (ARHAI) and the Rapid Review Panel (RRP) (on innovations in hospital infection control). His areas of specialist interest are the prevention and control of healthcare-associated infections, antibiotic therapy and antimicrobial resistance.

Professor Lawrence Paul Garrod

(1895–1979) was Professor of Bacteriology at St Bartholomew's Hospital until 1961, moving to the Royal Postgraduate Medical School as Honorary Consultant in Chemotherapy from 1965 until his retirement in 1971. He was co-author of the first five editions of the influential textbook, *Antibiotic and Chemotherapy* (1963, with Mary Barber; later co-edited with Francis O'Grady and Harold P Lambert; and from the 4th edn, containing a chapter on laboratory methods by Pamela M Waterworth). See Lambert (1976): 1.

Professor Curtis Gemmell

FRCPath (b. 1941) graduated in Bacteriology, University of Glasgow in 1963, and completed his PhD there in 1968. He was appointed Lecturer at the University of Glasgow, later Senior Lecturer, Reader and Professor in Bacterial Infection and Epidemiology (2000–06). He was visiting Professor at the University of Minnesota, USA (1979/80) and Director of the Scottish Methicillin Resistant *Staphylococcus aureus* (MRSA) Reference Laboratory (1997–2006).

Professor Alan Glynn

FRCP FRCPath (b. 1923) qualified at University College Hospital, London. He trained in clinical medicine but after two years as a senior registrar at St Mary's Hospital, London, he converted to bacteriology becoming Professor there in 1971 and Head of the Department of Bacteriology in 1974. In 1980 he became Director

of the Central Public Health Laboratory at Colindale, London until his retirement in 1988.

Dr Ian Gould

PhD FRCP(E) FRCPath (b. 1953) qualified in medicine in 1976 and trained in clinical microbiology and infectious diseases in the UK, Canada and Africa. He has been Consultant Clinical Microbiologist at Aberdeen Royal Infirmary since 1986, and Honorary Professor of Public Health, Epidemiology and Microbiology at the University of Trnava. He has been an adviser on antibiotic resistance and prescribing to the UK Department of Health, the Alliance for the Prudent Use of Antibiotics, the International Organisation for Epizoonosis, the European Commission, the European Centre for Disease Control and Government agencies abroad; and co-ordinator of the European projects ESAR and ARPAC. See Gould (2005).

Professor David Greenwood

(b. 1935) worked as a laboratory technician in various London hospitals, graduating in microbiology from UCL, London, in 1971, gaining a PhD in 1974. Later that year he moved to Nottingham as Lecturer in the new Medical School and was awarded a chair in 1989 as Professor of Antimicrobial Science at the University of Nottingham, later Emeritus. His research interests encompassed various aspects of the response of bacteria to antimicrobial agents.

Professor Jeremy Hamilton-Miller

PhD DSc FRCPath (b. 1938) was a microbiologist at the Department of Medical Microbiology, The Royal College Free School of Medicine (Royal Free and University College Medical School after 1998) between 1972 and 2003. He was appointed *ad hominem* Professor in 1987, later Emeritus. He had previously worked at medical schools at Guy's Hospital and Charing Cross Hospital, and at The Sir William Dunn School of Pathology. His main research was done on assessment of novel antibiotics, antibiotic resistance, natural products and urinary infections. He retains an interest in probiotics.

Professor Ronald Hare

MD (1899–1986), a colleague of Fleming at St Mary's Hospital, London, from 1925–30, was largely responsible for planning and building the penicillin plant at the University of Toronto funded by the Canadian Government. He returned to London as Professor of Bacteriology in the University of London at St Thomas' Hospital, London, in 1946 until his retirement in 1964, later Emeritus.

He was a member of the Council of the Wright-Fleming Institute (1952–60); the Nuffield Institute of Comparative Medicine (1960–68); President of the Pathology Section of the Royal Society of Medicine (1963/4) and member of the Council, Royal Society of Medicine (1965–68); as well as an Examiner in Universities of London, Malaya, Birmingham, West Indies and East Africa, Ibadan.

Dr Patricia Jevons

(1921–2005) found the first methicillin-resistant strain of *Staphylococcus aureus* at Colindale in 1960. See Jevons (1961).

Dr Angela Kearns

(b. 1959) trained as a microbiologist at the Public Health Laboratory in Newcastle General Hospital following her appointment in 1981. In the mid-1990s she set up a Regional Molecular Diagnostic Facility and played a key role in developing and implementing real-time molecular diagnostic techniques for a range of bacterial and viral pathogens now used throughout the PHLS. In June 2002 she transferred to the Central Public Health Laboratory (now known as the Centre for Infections, Health Protection Agency) in Colindale, London where she heads the national Staphylococcus Reference Unit and takes a keen interest in the evolution, epidemiology and pathogenicity of *Staphylococcus aureus*.

Dr Edward Lowbury

OBE FRSL (1913–2007) bacteriologist and poet, qualified in medicine at University College, Oxford, and trained at the Royal London Hospital, London. He served as pathologist in the RAMC (1943–46) and was a member of the scientific staff of the MRC Common Cold Research Unit, Salisbury, from 1947–49, moving to the MRC Burns Unit at the Birmingham Accident Hospital, as Head of Bacteriology, until his retirement in 1979. His particular interest was the prevention of infection; and the use of the mechanisms and the emergence of antibiotic resistance and its prevention. He was the first Honorary Director of the Hospital Infection Research Laboratory (HIRL) at Summerfield Hospital, Dudley Road (now the City Hospital), Birmingham, directed by Professor Graham Ayliffe. His work included Lowbury and Ayliffe (1974) and Lowbury *et al.* (eds) (1975). See also Ayliffe (2007); www.dudleyroad.org/History/History.htm (visited 14 August 2007). His reminiscences to Dr Tilli Tansey, dated 14 April 1998, have been deposited with the tapes, correspondence and other documentation from the 1998 Witness Seminar, 'Post penicillin Antibiotics', in GC/253 in Archives and Manuscripts, Wellcome Library, London.

Dr John Nayler

FRSC (1927–1993) joined the newly formed Chemistry Department at Beecham Research Laboratories in 1948, and was Head of the Department of Organic Chemistry there from *c.* 1960 to 1989. Nayler and his colleagues' demonstration of the existence of the penicillin nucleus (6-aminopenicillanic acid) in certain penicillin fermentation solutions led to the synthesis of most of the commercially and clinically important semisynthetic penicillins marketed by Beecham from 1959 to 1972. Nayler's name appears on the majority of the many patents and publications during this period. Doyle (1993). See also Batchelor *et al.* (1959).

Dr Bill Newsom

MD FRCPath (b. 1932) was Consultant Microbiologist to Addenbrooke's and Papworth Hospitals in Cambridge. He was on the editorial Board of the *Journal of Antimicrobial Chemotherapy*, and remains an Editor of the *Journal of Hospital Infection*. He has been President of the Hospital Infection Society and of the Institute of

Decontamination Sciences. He is currently the author of papers on medical history for the *Journal of Hospital Infection* and the *British Journal of Infection Control.*

Dr Marler Thomas Parker

FRCPath (1912–2006) qualified at Cambridge in 1937 and gained the Diploma of Bacteriology (London) with distinction in 1939 from the London School of Hygiene and Tropical Medicine under the tutelage of W W C Topley and G S Wilson. He served in the Royal Army Medical Corps as a pathologist, initially in the UK and then in India and Burma with the rank of Major. On his return to England he joined the newly formed Public Health Laboratory Service in 1946 and was Director of the Camarthen Laboratory until 1948 when he took over the directorship of the Manchester laboratory where he remained until he was appointed as Director of the PHLS Cross Infection Reference Laboratory (CIRL) from 1961 to 1978. He brought together the Staphylococcus and Streptococcus Reference Laboratories and introduced work on *Pseudomonas* and other emerging opportunist Gram-negatives. He carried out a number of seminal studies on the epidemiology of *Staphylococcus aureus*, particularly the characterization of the 80/81

epidemic strain and later with Dr Pat Jevons on the identification of 'Celbenin'-resistant microarray technology strains which are today known as MRSA. He was a Founder Fellow of the Royal College of Pathologists in 1964 and President of the Hospital Infection Society from 1984 to 1988. See Pitt (2006).

Professor Ian Phillips

(b. 1936) qualified in medicine at Cambridge in 1961 after studies at St John's College and at St Thomas' Hospital Medical School, London. His subsequent training was at St Thomas' Hospital and at Makerere University, Uganda, and in 1974 he was appointed Professor of Microbiology at St Thomas' Hospital Medical School. In later years he became Chairman of the British Society for Antimicrobial Chemotherapy, Chairman of the Association of Medical Microbiologists and finally, President of the European Society of Clinical Microbiology and Infectious Diseases. He became Emeritus Professor on his retirement in 1996.

Dr Tyrone Pitt

PhD (b. 1948) joined the Public Health Laboratory Service in 1965 and served in the Staphylococcus Reference Laboratory from 1971 to 1973. He was appointed Head of the Gram Negative Unit in 1981; this was combined with the staphylococcal service to form the Epidemiology Typing Unit and he has directed the joint service since 1994.

Dr Elizabeth Price

FRCPath (b. 1944) was appointed Senior Lecturer at the Institute of Child Health and Honorary Consultant Medical Microbiologist at the Queen Elizabeth Hospital for Children, London, in 1977. From 1991 to 1997 she also worked at Great Ormond Street Hospital for Children NHS Trust. In 1997, she transferred to the Royal London Hospital and retired from routine clinical work in 2006.

Professor Sir Mark Richmond

Kt PhD DSc FRS (b. 1931) graduated in Biochemistry at Cambridge and did three years' postgraduate work there under Professor Ernest Frederick Gale. His first postdoctoral position was at the National Institute of Medical Research, Mill Hill, and was followed by an appointment as Reader in Molecular Biology in the University of Edinburgh under Professors Martin Rivers Pollock and William Hayes. He moved to Bristol University in 1968 as Professor of Bacteriology, where most of his work on staphylococcal plasmids and antibiotic resistance

was carried out. In 1981 he ceased active research in microbiology on appointment as Vice Chancellor of the University of Manchester, during which term he became Chairman of the Committee of Vice-Chancellors and Principals of the UK. From there he became Chairman of the Science and Engineering Research Council. In 1993 he moved into industry as Global Head of Research for Glaxo. On retirement from Glaxo Wellcome in 1996 he became an Honorary Fellow in the School of Public Policy at UCL. Since retirement from full-time work, he has been appointed Non-Executive Director of a number of companies, notably Genentech Inc. and OSI Pharmaceuticals in the US and Ark Therapeutics in the UK. He has been awarded the Robert Koch Medal of the Robert Koch Stiftung in Germany, the Colworth Medal of the Biochemical Society and the Garrod Medal of the Society for Antimicrobial Chemotherapy.

Dr George Rolinson

(b. 1926) was Associate Director of Research and Senior Microbiologist at Beecham Pharmaceuticals at Brockham Park, Betchworth, Surrey, from 1955 to 1988. See Rolinson (1998).

Dr Geoff Scott

(b. 1948) has been Consultant

Clinical Microbiologist at University College London Hospitals since 1986. He was the first Press Officer for the Hospital Infection Society in 1987. He has an interest in hospital-acquired infections and manages patients with tuberculosis.

Dr Joseph Selkon

TD FRCPath (b. 1928) graduated MBChB in 1950 from Cape Town and DCP from London in 1954. He was Consultant Microbiologist and Director of the Newcastle Regional Public Health Laboratory from 1977 and then the Oxford Regional Public Health Laboratory from 1982. He has been an Honorary Lecturer at Newcastle and Oxford Universities and is a Past President of the British Thoracic Society (1988–93).

Dr David Shanson

FRCPath (b. 1944) was Senior Lecturer and Consultant Microbiologist at the London Hospital (1974–76), St Stephen's and Westminster, Charing Cross and Chelsea and Westminster Hospitals (1977–94) and was recently Microbiologist at Great Ormond Street Hospital for Children, London. He has been President of the Section of Pathology at the Royal Society of Medicine. In 1979 he chaired a steering group that founded the Hospital Infection Society and became its first Secretary in 1980, and later its Chairman (1984–87). In 1987 he chaired the organizing committee of the first International Conference of the Hospital Infection Society when MRSA was discussed as a major topic. He has been a member of all the UK joint working parties producing guidelines for the control of MRSA. See Shanson (1982).

Professor Reginald A Shooter

(b. 1916) formerly Professor of Bacteriology at St Bartholomew's Hospital, London, and an expert on hospital infection.

Dr Norman Simmons

CBE FRCPath FIFST (b. 1933) was consultant and clinical tutor at the Enfield Group of Hospitals for six years before being appointed Consultant Clinical Microbiologist and head of the Department of Clinical Bacteriology and Virology at Guy's Hospital London in 1972. He was responsible for the diagnostic and infection control services at that hospital until he left in 1994, but he continued in clinical practice until 2006. His special interests were endocarditis and food safety and he was a member of the government's Advisory Committee on the Microbiological Safety of Food from 1990 to 2001. He was appointed CBE in 2000. He retains his interests in hospital-acquired infection and food safety and still contributes to the literature on these subjects.

Professor Dale Smith

PhD (b. 1951) received his doctorate in the history of medicine from the University of Minnesota in 1979 under the direction of Professor Leonard G Wilson and was then appointed to the faculty there. He has been at the F Edward Hébert School of Medicine, Uniformed Services University of the Health Sciences, Bethesda, Maryland since 1982, Professor and Chairman of the Department of Medical History since 1997. He is the author of numerous papers on medical history. His professional interests include the history of graduate medical education, infectious diseases, surgery and the problems of patient evacuation in military operations. See Budd (1984).

Professor Brian Spratt

FMedSci FRS (b. 1947) was a postdoctoral fellow with Art Pardee at Princeton University in the 1970s where he discovered bacterial penicillin-binding proteins (PBPs) [Spratt and Pardee (1975)] and determined their role in the killing action of penicillin. He subsequently worked on

the mechanisms of resistance to penicillin mediated by alterations of PBPs and developed multilocus sequence typing (MLST) for the precise characterization of strains of bacterial pathogens. With Mark Enright he used MLST to identify and discern the evolutionary origins of the major lineages of MRSA. He was Professor of Biology at Oxford (1997–2001) and Sussex Universities (1989–97) and has been Professor of Molecular Microbiology and Head of the Department of Infectious Disease Epidemiology at Imperial College London since 2001.

Professor Gordon Stewart

(b. 1919) began his long professional interest in the control of communicable diseases with trials of domestic penicillins at the Royal Navy's Medical School, Clevedon, Somerset, in the combined Services Hospital, Trincomalee, Sri Lanka, and, from 1946, at St Mary's Hospital, London, other centres in the UK and the University of North Carolina at Chapel Hill, NC, from 1963 to 1972. He held the Mechan Chair of Public Health at the University of Glasgow from 1972 until his retirement in 1984, later Emeritus. He was particularly interested in the molecular aspects of antimicrobial and allergenic properties of penicillins, and contributed to efforts to widen activity and identify hazards in other antimicrobials and from other pathogens, like MRSA and Gram-negative bacteria. See, for example, Stewart (1965, 1992).

Dr Robert Sutherland

DSc (b. 1930) graduated in chemistry and microbiology in 1950 from the University of Edinburgh and joined the Chemotherapy Department of May & Baker in 1955 to evaluate novel macrolide antibiotics, including spiramycin. He later became head of the Bacteriology Laboratory of Beecham Research Laboratories at Brockham Park, Betchworth, Surrey, in 1962, involved primarily in research and development of semi-synthetic penicillins, resistance to β-lactam antibiotics and β-lactamase inhibitors.

Professor Sir Michael Swann

FRS (Lord Swann from 1981) (1920–90), Principal and Vice-Chancellor of the University of Edinburgh, 1965–73. See Mitchison (1991); Datta (1969).

Dr Tilli Tansey

HonFRCP FMedSci (b. 1953) is Convenor of the History of Twentieth Century Medicine Group and Reader in the History of Modern Medical Sciences at the Wellcome Trust Centre for the History of Medicine at UCL.

Mr Anthony (Phil) Tucker

(1924–98) graduated in fine art from Manchester College of Art and joined the *Guardian's* Manchester offices in 1953, becoming a sub-editor in 1957, a founder member of the features department and followed John Maddox as Science correspondent in 1964, where he was an articulate science writer, winning the Glaxo prize for science journalism several times, until his retirement in 1988. He continued until his death to write obituaries, such as Tucker (2006). See Radford (1998).

Miss Pamela Waterworth

(1920–2004) was a trainee nurse during the Second World War when deafness changed her career. She worked with Professor L P Garrod at St Bartholomew's Hospital, London, as a technician from penicillin's introduction in 1944 until Garrod's retirement in 1961. She was to join Mary Barber at the Royal Postgraduate Medical School, Hammersmith Hospital, London, but Barber's sudden death in a car accident prevented this. In 1971 she moved to the Department of Microbiology at University College Hospital, London, where she worked with Dr Joan Stokes until her retirement in 1981. Between 1955 and her retirement she published 69 peer-reviewed articles. Along with Joan

Stokes, she developed a method of qualitative antimicrobial susceptibility testing. See Ridgway and Stokes (2005).

Professor John West

MD FRCP (b. 1928) received his medical training in Adelaide, Australia, and then moved to the Royal Postgraduate Medical School, Hammersmith Hospital, London, where he spent 15 years. He has been Professor of Medicine and Physiology at the University of California, San Diego since 1969.

Sir Robert Evan Owen Williams

Kt (1916–2003), bacteriologist, trained at University College Hospital, London, and qualified in 1940. In 1942 he was appointed pathologist in the MRC Burns Research Unit at the Birmingham Accident Hospital and his interest in the spread of wound infections

began. He joined the Public Health Laboratory Service in Colindale in 1946 and was Director of the Streptococcus, Staphylococcus and Air Hygiene Laboratory from 1949 until 1960. He returned to the clinical field as Professor of Bacteriology and Director of the Wright-Fleming Institute at St Mary's Hospital Medical School, London, in 1967 becoming the first full-time academic to be Dean of the Medical School. He was appointed Director of the Public Health Laboratory Service from 1973 until his retirement in 1981. He was a member of the Medical Research Council and President of the Royal College of Pathologists (1975–78). See Figure 3.

Glossary*

automated laboratory techniques for detecting antimicrobial resistance
Standardized laboratory techniques on bacterial isolates were implemented in the 1950s, followed by clinical and economic pressures to find rapid test methods requiring less skilled labour input, which led to the development of semi-automated and automated antimicrobial susceptibility tests (AST) in the 1970s. Three routine ASTs are found in the UK: the disc diffusion test, the breakpoint test, and the **polymerase chain reaction (PCR)** test. Tests vary by typability, reproducibility, the ability to discriminate between strains, stability, cost-effectiveness and turnaround time.

bacteriophage type, group III
MRSA and other members of this group are often resistant to penicillin, streptomycin and tetracycline (PST). See Appendix 2, page 82.

bacteriophage typing
A biological typing method based on the susceptibility of different strains of *Staph. aureus* to different bacteriophages (bacterial viruses). Standard drops of a number of different 'phages are placed on a nutrient agar plate already inoculated with the bacterial isolate to establish the **strain**. 'Phage typing used in the 1950s was set out by Fisk [(1942)], who showed that many strains of *Staph. aureus* carried bacteriophages that could lyse or destroy some strains, but not all. See Williams *et al.* (1960): 26; Tarr (1958), see also Figures 2 and 13.

chlorhexidine
A dermal antiseptic used in gel and powder form to treat nasal carriers of MRSA, but with limited success.

cloxacillin (*Orbenin®*, Beecham Pharmaceuticals)
An oral **penicillinase-resistant penicillin,** a variant of oxacillin with an extra chlorine atom, introduced in 1962; it is used to treat infections with staphylococci resistance to penicillin.

colonization by MRSA
The spread of bacteria from one location, such as the nose, to the skin and from there to surfaces and other people. Colonization becomes a serious problem in a hospital setting with plastic devices and textiles to which the bacteria can easily spread. The majority of patients from whom **MRSA** strains

*Terms in bold appear in the Glossary as separate entries

are isolated are already colonized by the bacteria. See HPA (2007b): 6.

commensal

Normal flora; a harmless microbial association with a healthy host, who is asymptomatic.

daptomycin (*Cubicin*®; Novartis Europharm Ltd)

An acid lipopeptide antibiotic derived from *Streptomyces roseosporus*, a class of antimicrobial agents known as the peptolides, active against **MRSA** and **vancomycin-resistant Gram-positive bacteria,** including enterococci. European Union approval was received in 2006. The antibiotic has been available in the UK for treatment of bacteraemia and right-sided infective endocarditis since 2007. See www.nelm.nhs.uk/Record%20Viewing/vR.aspx?id=584977 (visited 8 October 2007). For nomenclature, see Filip and Cavelier (2004).

electrophoresis

A technique using an electrical field to separate a mixture of molecules by their differential migration through a gel or on specially prepared paper.

epidemic MRSA (EMRSA)

Strains with *mecA* genes and a property described as 'epidemigenicity' or 'epidemicity'. See pages 24 and 34.

epimers

Variants of a molecule with the same chemical structure but in a different steric arrangement, for example, differing only in the relative position of hydrogen and hydroxyl groups.

episomal or extrachromosomal

Refers to a genetic element separate from the chromosome.

European Antimicrobial Resistance Surveillance System (EARSS)

A European epidemiological network devoted to antibiotic-resistant pathogens. It provides reference data on antimicrobial resistance and is funded by the European Commission's Directorate-General for Health and Consumer Affairs (DG SANCO). See www.rivm.nl/earss/ (visited 15 January 2008). See also Figure 7.

flucloxacillin (*Floxapen*®, Beecham Pharmaceuticals)

A semi-synthetic penicillin, a derivative of 6-APA, containing one chlorine and one fluorine atom, better absorbed from the gut. It replaced **cloxacillin** in 1970 and is used to treat penicillin-resistant staphylococci.

glycopeptides

Antimicrobials, such as **vancomycin** and teicoplanin, that inhibit the synthesis of the bacterial wall and are used to treat **Gram-positive** infections including multi-resistant staphylococci.

Gram-positive bacteria

An organism that stains purple when treated with Gram's stain. The Gram-negatives stain red, a property thought to be associated with possession of a second, outer membrane in addition to the normal cytoplasmic membrane. See Health Protection Agency (2007a).

Health Protection Agency (HPA)

A non-departmental public body, independent of the UK Department of Health, created to provide support and advice to the NHS, local authorities, emergency services, and other agencies, as well as the Department of Health and the Devolved Administrations. Radiation protection became part of the HPA remit in 2005, replacing the National Radiological Protection Board. Its three major centres are the Centre for Infections at Colindale, the base for communicable disease surveillance and specialist microbiology; the Centre for Radiation, Chemical and Environmental Hazards, based at Chilton; and the Centre for Emergency Preparedness and Response at Porton, focusing on applied microbiological research and emergency response. For details of their current structure, see

www.hpa.org.uk/default.htm (visited 11 February 2008).

infection, MRSA

A minority of patients from whom MRSA **strains** are isolated are infected by the **MRSA**, although the proportion of colonized patients who later become infected varies between 5 and 60 per cent, depending on the population studied. See HPA (2007b): 6.

linezolid (*Zyvox*®, Pharmacia)

An oxazolidinone antimicrobial approved by the UK MHRA [Medicines and Healthcare Products Regulatory Agency] in January 2001, sometimes used for the treatment of **MRSA** and **vancomycin**-resistant enterococci.

mecA gene

The gene that confers methicillin resistance to *Staphylococcus aureus*. It can be detected by the polymerase chain reaction (PCR) method.

methicillin (BRL 1241; *Celbenin*®, Beecham Pharmaceuticals)

A β-lactam antibiotic first introduced by the Beecham Research Laboratories in 1960 and discontinued in 1993. It was suitable for treating infections caused by penicillin-resistant staphylococci because it was resistant to staphylococcal β-lactamase, but not to acid, and thus could not be given orally. See Barber (1961): 385. The trade name, *Celbenin*, was formed from the name of C E L Bencard. See Tansey and Reynolds (eds) (2000): 31.

molecular methods for typing *Staph. aureus* strains

Molecular typing methods include pulsed field gel **electrophoresis** (PFGE) and **PCR** ribotyping. For an evaluation of these methods on UK strains, see Hookey *et al.* (1998).

MRSA (methicillin-resistant/ multidrug-resistant *Staphylococcus aureus*)

Penicillin-resistant staphylococci were isolated with increasing frequency in the 1950s following widespread clinical usage of penicillin G [benzyl penicillin] and resistance was shown to be due to the production of β-lactamase, an enzyme that inactivates the β-lactam molecule. **Methicillin** is stable to staphylococcal β-lactamase and shows marked activity against most β-lactamase-producing staphylococci, but methicillin-resistant strains (MRSA) possess a modified **penicillin-binding protein** (PBP2a or PBP2′) with reduced affinity to methicillin or other β-lactams. The modified PBP is synthesized by a chromosomally acquired gene (*mecA*), located on a transposon, which enables spread between staphylococcal species. The transition from 'methicillin-resistant' to 'multi-resistant' has been gradual and is still incomplete. This term has continued to be widely used since methicillin became obsolete.

MRSA test

See **automated laboratory techniques for detecting antimicrobial resistance**

mupirocin (pseudomonic acid; *Bactroban*®; Beecham Pharmaceuticals)

An antibacterial preparation that received a UK marketing authorization (product licence) in 2001 and has been found to have remarkable activity as a topical treatment suitable for nasal carriage and skin infections due to staphylococci, as it is unrelated to other antibiotics and thus has no cross-resistance. See Hill *et al.* (1988); Casewell and Hill (1991).

neomycin

An aminoglycoside widely used during the 1950s, now considered too toxic for systemic use and restricted to topical application.

nosocomial infection

A hospital-acquired infection.

Panton–Valentine leuc[k]ocidin (PVL)

A major virulence factor of *Staphylococcus aureus,* whose

action impairs the respiratory activity of leucocytes, leading to their destruction. First described by P N Panton, Director of the Hale Clinical Laboratories, London Hospital, and F C O Valentine his Assistant Director. See Panton and Valentine (1932).

penicillinase-resistant penicillins

Penicillins, such as **methicillin** and **cloxacillin,** that are resistant to hydrolysis by staphylococcal penicillinase.

penicillin-binding proteins (PBPs)

The enzymes that carry out the final stages of cell wall (peptidoglycan) synthesis are inhibited by penicillin, resulting in a number of effects, including cell lysis and death. They were identified in 1975 as the physiological targets of penicillin and other lactam antibiotics. MRSA strains make an additional PBP that allows the bacteria to evade the action of β-lactam antibiotics. See Spratt (1975).

penicillins, isoxazolyl

Derivatives of 6-APA, including oxacillin (the first of the isoxazolyls), cloxacillin, dicloxacillin and flucloxacillin.

penicillins, quinoline-associated

For example, quinacillin (Boots), approved in 1961 and never used therapeutically.

'phage typing

See **bacteriophage typing.**

polymerase chain reaction (PCR)

A fast technique for making an unlimited number of copies of any piece of DNA and used as a procedure to detect the *mecA* gene in the chromosomal DNA of an MRSA sample. For the background to this 1986 discovery for which Kary Mullis shared the 1993 Nobel Prize for Chemistry, see http://nobelprize.org/chemistry/laureates/1993/mullis-lecture.html (visited 19 February 2008).

pre-admission screening

Some hospitals have introduced a clinic for patients with a planned admission to attend in advance for blood tests, an ECG and a chest X-ray, as well as an MRSA test, which involves a nose and sometimes a perineal swab, the results of which are available after 24–48 hours. Patients testing positive with MRSA are given antibacterial soap and an ointment to reduce nasal carriage for use at home and are screened again before admission. Emergency patients are assumed to be carriers and tested on admission, automatically given the treatment and isolated from planned admissions. For a description of NHS patient screening and national targets, see Department of Health (2008): 10–13.

quinupristin-dalfopristin (*Synercid*®; May & Baker)

An expensive injectible combination of two streptogramin agents (30:70 ratio) that received a product licence in 1999 for the treatment of nosocomial pneumonia, skin and soft tissue infections when there are no other antibacterial agents active against the organisms. It is a derivative of Pristinamycin. See Lamb *et al.* (1999).

resistance, multiple drug

Some **MRSA** strains demonstrate resistance to as many as 20 antimicrobial compounds, including antiseptics and disinfectants. See HPA (2007): 6.

scalded skin syndrome

A staphylococcal disease caused by strains that produce an epidermolytic toxin, one that necroses the skin.

staphylococcal bacteriophages

Viruses that lyse (destroy) specific strains of *Staphylococcus aureus.* A set of different bacteriophages (see Appendix 2) is applied to strains of staphylococci and varying patterns of lyses are obtained with different **strains**, for example 80/81, 6/7/53/75/77. 'Phage type 80/81 is a particularly virulent penicillin-resistant clone of *Staph. aureus* that has caused serious hospital-acquired and community-acquired infections, which was largely eliminated by the use of penicillinase-resistant β-lactamase antibiotics.

Staphylococcus aureus

A bacterium that may be found in nasal membranes, skin, hair follicles and perineum of healthy people and can cause superficial skin infections, boils and wound infection, but also, more rarely, pneumonia, meningitis, endocarditis and septicaemia. Its resistance to penicillin was described by Kirby in 1944.

Staphylococcus Reference Laboratory

Formed in 1949 as the Staphylococcus and Streptococcus Reference Laboratory by Dr R E O Williams. Currently it is part of the Laboratory of Healthcare Associated Infection, Centre for Infections, Colindale, London, of the **Health Protection Agency**. For a manual of services, see www.hpa. org.uk/cfi/dhcaiar/DHCAIAR_ User_Manual.pdf (visited 19 February 2008).

strain

A characterized pure culture of bacteria descended from a single isolate.

vancomycin (*Vancocin*®; Lilly)
A bactericidal antibiotic, whose
name was derived from the verb 'to
vanquish', isolated in the Eli Lilly
Company laboratories in the US
from a *Streptomyces* species found
in soils obtained from Borneo and
India, became available in 1956.
Lilly were sole suppliers of the drug
until 1983. Vancomycin-resistant
Staphylococcus aureus (VRSA)
have been described since 1995.
See Fairbrother and Williams
(1956); Kirst *et al.* (1998): 1303,
Table 1 and Figure 1. See also the
Supplement to *Reviews of Infectious
Diseases* [(1981) **3**: S199–300]
entitled 'Reassessments of
vancomycin – a potentially
useful antibiotic'; Griffith (1981);
Figure 11.

Index: Subject

Index: Names

Biographical notes appear in bold

Printed in the United States
120673LV00003B/69/P